MINES AND MINERS OF CORNWALL AND DEVON

For my wife Pip and her tin-streaming ancestors

MINES AND MINERS OF CORNWALL AND DEVON

THE TIN AND COPPER INDUSTRIES

ANTHONY BURTON

PEN & SWORD HISTORY

AN IMPRINT OF PEN & SWORD BOOKS LTD.
YORKSHIRE – PHILADELPHIA

First published in Great Britain in 2020
and reprinted in 2021, 2022 and 2024 by
Pen and Sword History
An imprint of
Pen & Sword Books Ltd
Yorkshire - Philadelphia

ISBN 978 1 52677 338 8

A CIP catalogue record for this book is available from the British Library.

Typeset in 10.5/13.5 pt Palatino
Typeset by SJmagic DESIGN SERVICES, India.
Printed in the UK on paper from a sustainable source by CPI Group (UK) Ltd, Croydon, CR0 4YY

Pen & Sword Books Ltd incorporates the Imprints of Pen & Sword Books Archaeology, Atlas, Aviation, Battleground, Discovery, Family History, History, Maritime, Military, Naval, Politics, Railways, Select, Transport, True Crime, Fiction, Frontline Books, Leo Cooper, Praetorian Press, Seaforth Publishing, Wharncliffe and White Owl.

For a complete list of Pen & Sword titles please contact

PEN & SWORD BOOKS LIMITED
47 Church Street, Barnsley, South Yorkshire, S70 2AS, England
E-mail: enquiries@pen-and-sword.co.uk
Website: www.pen-and-sword.co.uk

Or
PEN AND SWORD BOOKS
1950 Lawrence Rd, Havertown, PA 19083, USA
E-mail: Uspen-and-sword@casematepublishers.com
Website: www.penandswordbooks.com

CONTENTS

PREFACE

Like many others not directly engaged in the industry, my interest in the mines and miners of South West England was first aroused during holidays in Cornwall. There was a particular appeal for me, as the study of the development of steam power already fascinated me and the area was vital in that it saw the development of the first practical steam engine and, a century later, the design of the first railway locomotive. And I was also intrigued by the underground world of the miner. Over the years, I have visited mines of many kinds – I have crawled through Neolithic flint mines, filmed in a Roman gold mine, visited salt mines in France and many collieries, both here and in continental Europe. Talking to working miners, I have always been conscious of the pride they took in their work – work most of us would rather leave to others – and impressed by the resilience they and their predecessors have shown in the long struggle for safety and fair recompense for their work. All mines may seem superficially similar, but they are not, nor are all miners alike. The Cornish and their fellow workers have always been unique for their independence and noted for their skills – skills that have found a welcome around the world. Theirs is a very special story, so when John Scott-Morgan asked me if I would be interested in writing their history, I did not have to think about it, but agreed instantly.

A note on units. I have used those that were current at the time, which were the old imperial units: feet and inches; pounds and ounces; pints and gallons and so forth. There are good reasons for this. For example, an engine cylinder was bored as accurately as possible to a given size, which might be 40 inches and not 101.6 centimetres. Equally, the figures quoted for the amount of water pumped from a mine in an hour would be approximated usually quite roughly to thousands of gallons; it is difficult to convey the same result in litres. A fathom is six feet and again, distances have been left in the units which the miners would have thought in every day of their lives.

I am most grateful to Tony Brooks for reading the text and making many useful suggestions and corrections. Needless to say, any remaining errors are all my own.

Anthony Burton
Stroud

Chapter 1

THE ANCIENT WORLD

No one can visit Cornwall or West Devon without at some time or other catching sight of the distinctive shape of an engine house on the horizon. Such buildings suggest, by their sheer number, that they represent a once thriving industry, which indeed they do. They might also give the impression that this was an industry born in the Industrial Revolution and the age of the steam engine and that would be quite wrong. Tin, and to a lesser extent copper, have been mined in these areas for more than 4,000 years. The minerals have, of course, been in the ground for millions of years before man even appeared on the scene. In fact, it was around 200 million years ago that molten material from the magma at the centre of the earth was forced through earlier rocks, and crystallised as it cooled. The vapours from the magma, carried with them vaporised metallic compounds, that were forced up into cracks in the granite, where they too solidified on cooling. Unlike, say, coal, that was laid down in horizontal bands, the veins of copper and tin tended to appear as near vertical veins in the rock.

Mining is probably Britain's oldest industry. Even before metals were first used, the men of the New Stone Age were making tools and weapons out of flint. They found that the finest material was often under the ground in thick strata, so they began to dig down to extract it. At Grime's Graves in Norfolk, there is a huge Neolithic mine complex. If you go down one of these mines, you descend a shaft on ladders, and from the bottom of that, low, narrow tunnels lead away into the darkness. One doesn't even crawl through these, but more or less wriggle along on your stomach until the tunnel opens out into an area where the flint was excavated and even here, space is limited, with a roof scarcely more than three feet above you. The men who worked here had the simplest tools – deer antlers for pickaxes and shoulder blades for shovels. We know this because, from one of these pits, one can look through to an adjoining pit, where nothing has been disturbed for over 5,000 years and there are the tools they abandoned. A Cornish miner would recognise the different elements here: the shaft; the level running away from it; and the 'stope' where the valuable material was excavated.

Around 2500 BC, people moved out of the Stone Age and started using metal tools and weapons, by which time they had some five centuries and more of mining experience to draw on. The most important material was needed for an alloy that gave its name to this period – the Bronze Age. Copper is useful for many things, but it is quite soft and easily bent. However, it was discovered that if the molten metal is mixed with tin, in proportions varying from 5 to 20 per cent, it becomes harder and stiffer. No one really knows how this important alloy came to be discovered. Indeed, we do not even know how men first hit upon the idea of heating 'rocks' to produce metal. One theory is that stones used in hearths included some that were actually containing metal ores, and in the heat the molten metal appeared. This could only happen with metals that have a low melting point, such as tin at 232°C compared with iron that has a melting point of 1,538°C – which is why the Bronze Age lasted for nearly 2,000 years before the technology was available for smelting iron ore.

We have no written records for this long period of our history, but archaeology can tell us a lot and provide evidence that tin was indeed being mined in the South West at this time. The excavation of Bush Barrow, a burial site near Stonehenge, revealed a range of bronze objects, including a pair of daggers that were brought from Brittany, but the materials from which they were made were from England. The barrow is dated to c.2000BC, which suggests that Cornwall and Devon already had a thriving export trade in metals at this early date. Further confirmation appeared in 1999, when two German metal detectors unearthed a find in Saxony that included the remarkable Nebra Sky Disc. The disc itself was bronze, on which were gold figures representing the sun and moon together with a few stars and lines indicating the solstices. Analysis has shown that although the copper came from Austria, the tin was from Cornwall, and the disc has been dated to around 1600BC.

Cornish tradition has it that it was the Phoenecians who were the major traders with Cornwall but there is no archaeological evidence for this and the existence of Cornish tin in the Nebra disc shows that the trade started some centuries before the Phoenicians became a powerful force in the Mediterranean region. There are several accounts of trade with Cornwall, but all written long after the event. Greek historians wrote of the Cassiterides or 'Tin Islands' off the coast of Gaul. Strabo (c.64BC to 24AD) in his Geography described them as ten islands, one of which was uninhabited, but he had never actually been to the area himself. He makes no mention of tin trade. Diodorus Siculus, writing in the first century BC but apparently quoting earlier travellers, described the inhabitants of Cornwall as being very friendly, mainly because they were used to trading with foreigners. He describes how they mine and prepare the tin, casting it into blocks and then taking it to an island called Ictis, that can be reached by carts at low tide. From there it is loaded into boats and taken across the Channel and then overland by pack horse to the Mediterranean. No one has identified Ictis but the likeliest candidate is St. Michael's Mount.

One thing we lack is evidence of the techniques used to obtain the ore in prehistoric times. Centuries of working over the ground have obliterated any traces of early workings. Tin appears naturally as the ore Cassiterite, tin dioxide and was formed when granite pushed through earlier sedimentary rocks. It appears as veins that dip steeply, can be several kilometres long and vary in thickness from 0.5 to 3 metres. It was discovered in the first place because streams washed out some of the ore and brought it to the surface. The ore can be recognised easily, being much denser than other stones found with it. At the Blue Hills tin streaming works, near St. Agnes, they work with material collected from the nearby beach. Perhaps the earliest tin workers also collected material washed to the surface, but eventually it must have become obvious that the material had originally come from some richer source and by following the stream they could have found lodes close to the surface. How far they went underground is not known – Diodorus simply says they got the material from the earth, which suggests some form of mining.

The development of furnaces to smelt iron brought in a new age and bronze was no longer used for tools and weapons. The material did, however, still find a use for ornaments and other small objects, while copper would later be used for coinage. There is evidence that tin was still being mined in Cornwall around 300BC. Chun Castle is an Iron Age hill fort near Penzance and excavations in the 1920s discovered a series of small pits that had been used for smelting, in one of which was a large lump of slag, which proved to be from tin working. When the Romans arrived, new uses for tin were found. They brought with them a new material, pewter, an alloy that has a very high tin content, 85 per cent or more. It was mostly in the form of tableware and its use continued right through the Middle Ages among the wealthier classes. Sadly, little is known about the Roman trade in Cornish tin, and the paucity of Roman sites west of Exeter suggests that there was little impact on the region.

The period following the withdrawal of the Romans is generally referred to as the Dark Ages and when it comes to information about mining in the South West, the name is all too apt. We know very little about the amount of activity that was going on and there are no statistics on the quantity of metal produced. There are legends, including the arrival of Joseph of Arimathea to trade in Cornish tin. There are several references to Jews in the tin trade, but it is not clear that the name has the same meaning as it does today, and seems to have been applied more or less randomly to anyone from southern Europe and the Mediterranean. One more reliable source comes from Leontios of Neapolis' biography of John the Almsgiver, Patriarch of Alexandria at the beginning of the seventh century. He records a ship full of grain being sent to relieve a famine in Cornwall and returning with a large quantity of tin. Nothing else appears in written documents until the early medieval period but, following the Norman invasion, there was an increased demand for metal. More and grander churches were being built and required bells and bell metal is a form of bronze with a much higher proportion of tin than usual, as much as 20 per cent. But it is only in the twelfth century that we begin to get written accounts of the industry. The earliest of these is Richard Carew's *Survey of Cornwall* in 1601. There is a long, informative section on the tin industry, which begins by describing the Cornish belief that the tin was deposited during Noah's flood. Later, he moves onto more recent and reliable history, describing the regulation of the industry in the twelfth century. In the following extracts, the spelling has been modernised for clarity.

> 'After such time as the Jews by their extreme dealing had worn themselves first out of the love of the English inhabitants and afterwards out of the land itself, and so left the mines unwrought, it happened that certain gentlemen, being Lords of seven tithings in Blacknoore whose ground were being stored with the mineral grew desirous to renew this benefit: and so upon suit made to Edward, Earl of Cornwall, son to Richard, King of the Romans, he obtained from him a Charter with sundry privileges: amongst which it was granted them to keep a Court, and hold plea of all actions, life, limb, and land excepted: in consideration whereof, the said Lords accorded to pay the Earl a halfpenny for every pound of tin which should be wrought, and that for better answering this tax the said tib should be brought to certain places purposely appointed, and there pieced, coined, and kept, until the Earl's dues were satisfied.'

This was the start of a period which recognised the Cornish tin industry as having a unique status. This did not, however, apply to the local copper mines. Greater changes were soon to follow.

Richard I has been regarded as a national hero – the Lionheart – in contrast to dastardly King John, but the citizens of the day may have had a less exalted view of the monarch. His crusades cost the country a great deal of money and taxes had to be raised to pay for his overseas ventures. In 1196, Richard was off again leaving the job of running the country to Hubert Walter, who somehow managed to combine the roles of Lord Chancellor, Chief Justiciar and Archbishop of Canterbury. One area where he hoped to raise revenue was among the mines of Devon and Cornwall and he sent William de Wrotham to investigate and appointed him Warden of the Stanneries, giving him a measure of control over the tin mining industry. He found a very disappointing situation, discovering that only around 70 tons of tin were being sent out of the area in a year, which raised a paltry £17 in tax. The Cornish business in particular was languishing, largely because the work was being carried out piecemeal, often by part time miners working small holdings.

De Wrotham realised that the industry needed to be better organised and called together the leading tinners to discuss what needed to be done. He standardised the way in which tin was sold, by creating official marks for half-ton slabs. He increased the rate of taxation and was able to send more money to the hard pressed Lord Chancellor. What he did not do was attempt to improve production. But major changes were under way. When John came to the throne, he was as cash strapped as Richard had been but took positive action to improve his revenue and brought a major change to the Cornish industry. To understand what his reforms meant in practical terms, we need to look at the curious nature of the industry at that time.

From early days mining rights had depended on a system known as 'bounding'. This allowed prospectors to enter an area and claim a portion for development. In Devon, the copper miners had free access to all land but, in Cornwall, certain areas which belonged to the lords could only be entered with their permission. A Warden was appointed to try and resolve disputes. Large areas of the county were, however, free for anyone to enter. Carew described how the system worked.

> 'When a mine (or streamwork) is found in any such place the first discoverer aimeth how far it is likely to extend, and then at the four corners of his limited portion, diggeth up three turfs and the like (if he list) on the sides, which they term Bounding, and within that compass, every other man is restrained from searching.'

The problem was that often the workings were too small to be profitable and many who held tin bounds looked on them as a supplement to their normal incomes, perhaps working smallholdings. The bounds were not necessarily valuable. My wife has Cornish ancestry and one of her forbears, Melchisedech Rogers, born in 1668, had rights on Bodmin Moor. When he died, they were valued in his will as worth less than the family bed.

In order to regulate the trade and encourage entrepreneurs, a new Charter was issued under King John in 1201, in which the tinners were awarded the rights of 'digging tin, and turfs for smelting it, at all times, freely and peaceably and without hindrance from any man, everywhere in moors, and in the fees of bishops, abbots and counts, and of buying faggots to smelt the tin without waste of forest, and of diverting streams for their works, as by ancient usage they have been wont to do'. It was not just in their working lives that the miners were given a special status. They were no longer under the jurisdiction of a local magistrate, but instead answered to the Warden. It was almost as if they were a self-governing community within the wider world. This was a time when the majority of the poor were serfs, answerable to the lord of the manor. The Cornish miners were, in effect, free men. They were independent and that independence was to be a characteristic that they would carry on down through the ages.

The passing of the Charter had an immediate effect in increasing productivity and in the next chapter we will take a look at the actual nature of the workings in the late medieval and renaissance period.

Chapter 2

THE EARLY WORKINGS

In its earliest form, as practised in the Middle Ages, collecting ore was a comparatively simple process. The tin ore would be washed out of the veins by natural erosion. This was a process that took thousands of years, with the heavier tin ores settling in the valley. The ore particles, being heavier than the surrounding gravel, could easily be separated out by panning. This involves swirling the material round in a dish with water, so that gradually all the lightest material would be washed away, leaving the heavy, valuable ore. Until around 1600, all tin was acquired this way, mainly from Dartmoor. It must have been obvious that the material had washed out of the ground, and might appear at the surface, either in a sea cliff or on a hillside. To get it out of the ground required a new technology – mining. Once again, Carew provides us with a picture of the mining scene in the early years. His description starts by discussing the new form of extraction – streaming, developed into mining. The process starts when the prospectors find the typical heavy, smooth stones on the surface. They then start digging a trench away from that spot, typically 5 to 6 feet deep and 3 to 4 feet wide. Often it appears that the works will arrive at a stream. They then divert it into a new course to reveal the stream bed. Carew was somewhat disapproving:

> 'This yieldeth a speedy and gainful recompense to the adventurers of the search, but I hold it little benefit to the owners of the soil. For these low grounds before time fruitful from there having herethrough their wrong side turned outwards, accuse the Tinners injury by their succeeding barrenness.'

Carew then moves on to describe 'lode working'. The start of the process is similar to that for streaming, locating material, known to Carew as 'shoad' near or on the surface. This they believe will have been washed out of some richer source, so they must try and work out where the rich lode might be. Having decided on a likely direction, they move up the slope and dig a shaft, usually 5 to 6 feet long, 2 to 3 feet wide and 7 to 8 feet deep in the hope of reaching the lode. If that fails, they try a different direction and sink another pit and continue sinking pits until they either find a rich source of ore or abandon the area. Even when they do reach an ore bearing area, it has to be of the right sort:

> 'They also discover which was the quick ground (as they call it) that moveth with the flood and which the firm in which so much shoad doth lie.'

Ideally, in order to prove profitable, the lode should be around a foot and a half thick, which would be considered rich, but lodes of just a foot in thickness could also yield good profits if they turned out to be part of a more extensive system. It was a matter of fine judgement to decide when an enterprise should be followed through and when it should be abandoned. Carew spelled out the results of making the wrong decision:

> 'But you may not conceive that every likelihood doth ever prove a certainty oft divers have been hindered, through bestowing charges, by seeking and not finding, and many undone in finding and not speeding, whiles a fair show, tempting them to much cost, hath in the end failed in substance, and made the adventurer Bankrupt of their hope and purse.'

Carew goes on to mention a much easier way of finding riches – relying on revelations in dreams. He quotes examples, starting with a 'gentlewoman' who, in the reign of Edward VI, dreamed that a certain area was full of high quality ore, told her husband who acted on her word and made a fortune. William Pryce, writing in his *Mineralogia Cornubiensis* in 1776, mentions an alternative method of looking for the mineral – divination. Although he admits the subject is controversial – as indeed it still is – he devotes several pages to describing the techniques involved.

Discovering a good lode was only the beginning of the operation. The next stage was to start mining. In general, the prospector who had uncovered the lode then had to go into partnership with others to fund the development. Sometimes the partners would be working miners and sometimes employers, who hired labourers. The hired workers were paid around eight pence a day or, if taken on for a longer period, five to seven pounds a year. The work was overseen by the mine captain, who was responsible for all aspects of the work from timbering to allotting tasks to the men. Carew does not seem very impressed with the personnel, even while acknowledging the difficulty of the endeavour:

> 'In most places their toil is so extreme, as they cannot endure it above four hours in a day, but are succeeded by spells: the residue of the time they

> wear out at quoits, kayle [a type of skittles] and the like idle exercises. Their Calendar also alloweth them more Holy-days than are warranted by the Church or Laws, or their own profits.'

The men had only the most basic tools. The pickaxe had an iron point at one end and a flattened section at the other. The latter was used to drive iron wedges into cracks to break open the rock. The only other tool mentioned is a broad shovel. With these basic implements, they dug shafts down to the lode, and then worked outward from the bottom of the shaft – and Carew adds the intriguing note that in the deepest shafts, which could go down 59 fathoms, men could look up at noon and see the stars. The men continued working away from the shaft, until they reached a point where the air became too foul to breathe, at which point a new ventilation shaft had to be sunk. In places, the men met loose earth that had to be shored up with timbers, but it was always a dangerous area. There were reports of men being buried under falls or having their exit blocked and becoming trapped underground. Hard rock simply meant hard work and in some cases, a man might not advance more than a foot in a week. Carew described these men as playing moldwarps (the word in some areas for a mole). They were raised and lowered in the shaft by means of a stirrup at the end of a rope, wound by a two-man hand winch.

Water was the greatest enemy and was dealt with in a variety of different ways. If not too great an inundation then hand pumps could do the job, but otherwise simple bucket pumps were used, operated by a water wheel. If these measures did not work, then an adit had to be dug to drain away the water. These could be extensive – and we shall be looking shortly at the most complex system of them all created in the eighteenth century. Even in Carew's day, these were impressive works of engineering as he appreciated, even if he was less than flattering about the men who conceived and built them.

> 'If you did see how aptly they cast the ground, conveying the water by compassings and turnings to shun such hills and valleys as let them by their too much height or lowness you would wonder how such great skill could couch in so base a Cabin as their otherwise thick and clouded brains.'

Once the ore had been brought to the surface, it had to be treated to be made ready for smelting. The large rocks had to be broken up with hammers and then either loaded into carts or taken by packhorse to the stamping mill. It seems that carts were actually something of a rarity nevertheless, as if one appeared in Penzance in those days, people would come out of their houses to see it. The stamps themselves consisted of massive baulks of timber, tipped with iron. They were worked by a water wheel, with a simple mechanism, consisting of a rotating cylinder with projections. The timber stamp also had a small projection. As the cylinder turned, the two projecting pieces would meet and then clear, first lifting the stamps and then allowing them to fall back down to pulverise the ore. If further grinding was necessary, the material was sent to the 'crazing mill', which used water-powered grindstones.

The next stage was required to separate the heavy cassiterite from the other matter. The stream that had worked the wheel was led away over a series of ledges, each of which was covered by a piece of turf, usually about 3 feet by 4 feet wide and about one foot thick. Not any turf would do; it needed to have the right density of grass so that as the stream washed over it, the lighter waste material was carried away but the heavier ore was held. The best material was found near St. Michael's Mount, where the turf was dug from an area that was covered with water at high tide. It may have been necessary to add one final stage. The material was placed in a circular dish about two feet diameter, with a pair of handles. The worker put in the material together with some water and carefully swirled it round, allowing the lighter material to be washed over the edge of the dish. This process will be familiar to anyone who has ever seen film of prospectors panning for gold.

The next stage was to smelt the material in the blowing house. Carew has little technical information to offer on this, other than to say that the ore was smelted with charcoal, brought to a high temperature by water-powered bellows – hence the name blowing house. The molten tin was cast into blocks weighing from 300 to 400 pounds, and then each block was impressed with the owner's mark and sent to the Coinage. For once, Carew, who has generally been less than generous in his estimate of the abilities and intelligence of Cornish workers, seems actually to recognise the great efforts they make.

> 'I have already told you how great charge the Tinner undergoeth, before he can bring his ore to this last mill, whereto if you add his care and cost, in buying the wood for their service, in felling, framing and piling it to be burned, in fetching the same, where it is coaled through such far, foul, and cumbersome ways to the blowing house, together with the blowers two or three Months extreme and increasing labour, sweltering heat, danger of scalding their bodies, burning the houses, casting away the work and lastly their ugly countenances, tanned with smoke and besmeared with sweat: all these things (I say) being duly considered, I know not whether you would

> more marvel either whence a sufficient gain should arise to countervail so manifold expenses, or that any gain could train men to undertake such pains and peril.'

The reference to 'burning the houses' is later explained. Many of the early blowing houses had thatched roofs, and not surprisingly the sparks from the furnace might cause them to catch fire, but then it was discovered that there was valuable tin in the ashes, so it became the curious practice to deliberately set fire to the houses once every six or seven years, as the value of the tin was considerably greater than the cost of rebuilding. It seems the practice was already dying out in Carew's time and new blowing houses were being built with wide, sloping chimneys in which the tin particles could settle.

The work of tin streamers was described in more detail by Pryce in his *Mineralogia Cornubiensis* written over a century and a half after Carew. It is not always clear whether they are describing the same processes or ones that have changed and developed over the intervening period. Pryce does, however, provide a lot of details that help to fill out the picture of tin streaming over the years.

He begins by listing several ways in which the mineral might be located, some of which seem quite odd. For example, one way in which copper deposits could be found was by tasting the water in a stream, or by leaving a piece of iron in the stream for a few days to see if it acquired a green tinge. But the commonest methods used by tin streamers seems to have been much as Carew described them, with Pryce adding extra information. Some methods seem quite haphazard – simply digging a whole series of shallow pits in the hope of finding something useful. A much more common method followed the discovery of stones in a valley that contained ore. The tinners then had to try and work out where it might have originated. They assumed it had been washed away from a lode higher up the slope and set about testing the ground. An alternative method was developed because they knew that tin and copper lodes generally run in an east-west direction. Having decided on the likeliest area to be investigated, they would cut a number of drifts running north-south and in that way would be sure of cutting any lodes. Pryce records that he tried the method himself with great success.

In order to develop the site, they would have to reach an agreement with the landowner, who normally accepted payment in a proportion of the black tin – the pure but unsmelted ore. This could be anything from an eighth to a sixth of the total raised annually. There was also a cash payment to the lord for every man employed on the site, with a slightly smaller payment for boys. The method of working described differs from that mentioned in Carew, so one might assume that the technology had changed over that quite long period of time.

The work was begun by sinking an exploratory shaft from 3 to 7 fathoms deep until it hit a stratum of rock or clay, where tin is normally found. The streamer then took a shovelful of material, washed away the dirt and determined if there was good material that was worth working. Tin strata can be anything from one to ten feet thick and can extend a short way or reach the whole length of the valley, and the ore can vary in size from that of a walnut to fine sand. If the sample was satisfactory, the streamer went to a lower part of the valley to dig a 'level', a long trench that acted as an adit, drawing away water from the higher seam, and carrying with it material and waste. If the material seemed satisfactory, then work could begin, and the men both removed the material and began the separation process. Each shovelful from the stream was thrown onto a 'tye', a sloping board structure, 4 feet wide, 4 feet high and 9 feet long. Water ran down the board, and the excavated material was regularly turned. The waste was washed down but the heavier tin remained at the top of the board. It would be further cleaned, using a similar system, but worked more carefully and finally sieved. Large amounts of waste were inevitably removed and Pryce is critical of how it was usually treated,

> 'If there is a copious stream of water near at hand, they cast this refuse into it, by which it is carried so far as to make its exit into the sea, for which practice they certainly deserve our severest censure: at least, if the choking of harbours and rivers, and the destruction of thousands of acres of improvable meadow land, are not more than an equivalent for the casual and temporary profits arising from Stream Tin.'

Pryce goes on to show how the land could be restored after the stream workings had been completed and abandoned. He also mentions that the waste washed down to the sea might still contain some tin. A group, known as 'lappiors', which is a name that he suggests comes from the Cornish word for 'dancers', reclaimed these traces by working the material with their bare feet. He says that poor women and children regularly worked the land below Perran Arworthall, near Falmouth, which is covered by the sea at high tide. The returns were poor but for the local community, the effort was worthwhile.

Pryce also gives a more detailed account of the driving of adits, which had become a vital part of the whole system of draining mines. The adits were generally six feet high and two feet six wide. This allowed enough space for the men to work, who were driving the adit, and also enough room to remove the waste using wheel barrows. If the ground was very hard, then the adit had to be made larger to allow the men room to work. Generally the work required four men, working in shifts day and night and possibly one or two boys to wheel away the waste. Ventilation shafts had to be sunk at regular intervals, usually around 250 feet apart. These were generally 6 feet long by 4 feet wide and they required six men working day and night. Ventilation could be improved by putting in a 'sailer', which was a wooden floor set about one foot above the bottom of the adit and which stopped short of the shaft and drew the air through. Another option was what Pryce refers to as an air pipe, which appears to have some sort of vane at the top, turned by the wind at the surface.

The workers were paid by the length of adit dug. However, when the workers met particularly hard ground, they would try and find a way round it, which made their task easier but increased the length of the adit. This, of course, meant that their employers had to pay more for the work, which they considered unreasonable. As a result, they began to pay not by the length measured underground, but on the surface as a straight line. As described by Carew, removing water from an adit was essential, and a number of methods were used. The simplest was to scoop it up into barrels to take to the surface. These could then be lifted up the shaft by means of a horse gin or whim set at the top of the shaft. This consisted of a type of windlass operated by means of a horse walking round and round a circular track, harnessed to an overhead beam that turned the barrel of the windlass.

Work on adits could continue for years and Pryce himself was involved in one taking seventeen years but he declared that the effort was worth it. This was not the longest time ever taken. That honour goes to what is known as the County adit. Work was begun by the Williams family in 1748 in Gwennap and continued for fifty years, at the end of which it had extended to thirty miles and drained forty-nine mines. Pryce also wrote about the dangers of driving adits in wet mines.

> 'Whenever they are apprehensive of coming towards the house of water, as the miners term it, they bore a hole with an iron rod towards the water about a fathom or two or so many feet further than they have broke with the

> pick-axe. As they work on, they still keep the hole with the borrier before them that they may have timely notice of the bursting forth of the water, and so give it vent or passage. Yet not withstanding all this care and prudence, they are often lost by the sudden eruption of the water. In some places, especially where a new Adit is brought home to an old mine, they have unexpectedly holed to the house of water before they thought themselves near it, and have instantly perished. Some have driven by the side of the house of water and have perished also by its unexpected eruption.'

Jenkins, in his history of Cornish miners, tells the story of just such an accident in the St. Just area in the 1870s, as told to him by an old miner.

> 'I can mind the day the four men was drowned holing through into the house of water in North Levant. 'T'was a Monday and I was working afternoon core out Spearn. They say the end they was driving had been bone-dry all along. When the men got down that day they stopped out at the beginning of the level to touch a pipe of bacca, and told the boy to go in and clear up the end 'gainst they started to work. Over a while the boy came out and said: "We aint far off the water now, Uncle Nick, for tis running through in the end." "Git away," said the man, "theest took fear, booy." What happened after that they don't know, but 'tis proposed the men went in to work and the first blow they struck on the drill, the water burst through upon them. One man was found afterwards with the tram thrown on top of him yards back in the level, and one they never fond for a week. The two other men, as I said, I seed myself broft into the carpenters' shop and laid out 'pon the binch.'

One can only imagine the force of water that burst over the men, sufficient to throw one backwards and to throw a heavy cart over him. Such accidents were sadly not all that rare. But the adits were essential if the valuable metal was to be won from the ground.

Once the tin ore had been thoroughly cleaned, it was taken to the blowing house. This is very different from the thatched structures of Carew's day. Now the furnaces, known as 'castles', were constructed from massive stones, bound with iron. Blast was provided by a pair of pipes, through which air was blown from a pair of bellows worked by water wheels. The pipes passed through wrought iron plates set ten inches above the base of the furnace at the 'eye of the hearth'. The furnace was charged with alternating layers of ore and charcoal; it took 24 6-gallon packs of charcoal to smelt 12 cwt of tin. The molten metal was tapped through a hole, 4½ inches wide and run off into a large stone trough. From there it was ladled into smaller troughs that acted as standard moulds to produce blocks of the pure metal for sale. The tin from streaming was known as grain tin and commanded a high price.

Carew has a colourful account of selling the tin. Merchants would arrive from London and regale the locals with their tales of the parlous nature of trade, of ships lost at sea and how these outwardly wealthy men were really on the verge of pauperism. In such circumstances, they could hardly be expected to afford to offer very high prices. In return, the tinners described how difficult the works had proved that year, how the price of essentials had risen and so they must, most unfortunately expect a really good price. Having established their two extremes, the parties then got down to the serious business of bargaining and reaching the compromise price, which probably both parties had reckoned on in the first place.

From these sales, the Cornish had to pay a tax, which was far from popular, and there are many indications of tin being sold overseas and no transaction ever being officially recorded. A.K. Hamilton Jenkins in his book *The Cornish Miner* (1927) quotes a letter sent to the government offices in London in 1624 by the Ambassador in Constantinople, Sir Thomas Rowe:

> 'Whereas complaint hath been made to the Prince his Highness, by the farmers of the pre-emption of tin, that both His Majesty suffereth great loss in his customs, and that thy are very much hindered by the secret exportation of tin in slabs, moulds, and bars: whereof a great part is (as we are informed) brought to the ports of Constantinople, Aleppo, and Smirna; although we have taken the best course we can here for the remedying of this abuse; yet we … pray you take special care, that when any ships shall arrive in those ports, you cause diligent search to be made for such tin brought without the licence of the farmers, and if any shall be found, to seize the same, satisfying the mariners for it, in some reasonable way, but not so as that they may be encouraged to use the like fraud afterwards.'

Neither Carew nor Pryce have a great deal to say about the living and working conditions of the local workforce. Other writers are all agreed that the lowest class, those who worked for wages, were very poor. In the sixteenth century, wages could be as low as eightpence a week, and it was estimated that a single man needed at least tuppence to feed himself. How a married man with a family managed remains a mystery. Yet, although all were poor, they still looked after each other. If a man was sick and receiving no pay they would ensure that he and his family did not starve, but shared out their own meagre resources. Those workmen who had banded together to work a mine did not pay wages, but instead shared out the black tin itself. They looked after each other but took a less charitable view of any foreigners who might try mining on their own account. Once discovered, any tin they had dug was confiscated and they were marched off to gaol. Yet there are accounts of German miners being brought over, simply because they were acknowledged as experts in the field. One of the richest mine owners in Cornwall, Sir Francis Godolphin, brought over a German expert around 1580 to advise him and he is said to have built the first water-powered stamps to be used in the region. Not that the Cornish themselves lacked expertise. When, in 1607, silver deposits were discovered in Scotland, a request was sent out for twenty experienced Cornish miners to travel north to advise the Scots.

By the time Pryce wrote, streaming and bounding were becoming rare. One reason was that mining technology was moving on. But there was also the fact that anyone wanting to begin operations had to apply for permission to the Stannary Court, and that gave local landowners a chance to step in and claim prior rights. The whole industry was quite rigorously organised, and that will be the subject of the next chapter.

Chapter 3

THE STANNARIES

The Charter of 1201 had laid down the rights of the tinners, but it also divided up the tin bearing regions of Devon and Cornwall into stannaries. There were four in Cornwall, corresponding to the main ore-bearing regions: Penwith and Kerrier in the far west, extending down to Land's End; Tywarnhaile, covering the area from St. Agnes to Carn Brea; Blackmore, the region round St. Austell; and Foweymore, the old name for Bodmin Moor. Three stannaries were also established in Devon, usually known by their main towns: Tavistock; Ashburton; and Chagford. The Devon tinners succeeded in having Plympton approved by the High Sheriff as a fourth stannary town on the grounds that being near Plymouth it was more accessible to merchants arriving by sea. None of the different stannary areas were ever precisely defined.

The stannary towns had a very special importance, as they were home to the coinages, to which all refined tin produced in that particular stannary had to be brought for sale. In Cornwall these were Penzance, Truro, Helston, St. Austell, Liskeard and Lostwithiel. The thinking behind the coinages was that they were overseen by officials who could guarantee such important matters as the accurate weighing of the tin and could assess its quality. They could also, of course, make sure that all the taxes due on sales were paid. This was all very satisfactory for the officials, but not necessarily an advantage for the tinners. In the early days, the sales only took place twice a year, which meant that the tinner only got paid twice a year, and all too often, a tinner found himself in the position of not having enough money to carry on his business until the next coinage. He would then have little choice other than to borrow funds at an often exorbitant rate of interest. It was all too often the case that these unfortunate men found that after the coinage, when they had paid off the debts and interest, they were soon once again short of funds, and the whole cycle of debt was repeated.

The system did not always work in the merchants' favour either, as they had to have the money available to purchase large quantities, rather than buy a little at a time. And the only way they could have the funds available was to bring the money with them. This was not without its hazards. Travel by sea and you risked shipwreck, by no means unknown off the rugged Cornish coast. Go by land and you had to contend with wretched roads, dangerous river crossings and the risk of being attacked and robbed.

A system was introduced which attempted to get rid of the worst abuses of the system by purchasing the bulk of the tin for the Crown. In February 1606, Sir Richard Smith set off for Cornwall with £13,000, equivalent to more than a million and a half at today's prices. The money was carried on 26 horses and, not surprisingly, given the sum involved, there was an escorting party of sixteen men, each armed with a pistol. The whole trip took twenty days, for which the men were paid six shillings each. Sir Richard also had to pay for provisions and lodgings for the journey. It was a cumbersome way of doing business and it seems the idea of buying in such huge bulk was soon abandoned and the job of purchasing tin was left to the usual London merchants and the few that came across from continental Europe.

Once a sale was announced at a coinage, the town would begin to fill with all those who would be involved. The tinners would bring in their refined metal and would have to pay the tax of four shillings a hundredweight before sales could begin. The sale itself was a grand and very formal affair, attended by officers appointed by the Crown. Proceedings started in the coinage hall at noon and the crowds assembled. The merchants also gathered and began haggling with the tinners whose metal they planned to purchase. Once everyone had settled down, a space was roped off for the officials who would be in charge of the sale. The controller and receiver brought the stamping hammer to mark the tin blocks and the official weights. The weighers and assay master were there to accurately measure the quantity and purity of the metal. A special weighing machine, known as the King's beam, was brought in together with the weights, which had been sealed to ensure that no one had tampered with them since the last coinage. They were now handed to the weigher. The assay master then appeared with his hammer and chisel and the officials – the steward, the controller and the receiver – took their seats to oversee the procedures. When everyone was ready, porters brought in the heavy blocks and put them on the scales. Once the block had been weighed, the weigher shouted out the result and the officials noted them down. After the weight had been recorded, the block was taken off the scales and the assay master chipped off a small fragment and assayed it to confirm its quality.

The assay system was not unlike panning, but instead of a pan, a vanning shovel was used. The dried sample was weighed and then mixed with water and swished round on the shovel, washing away the lighter particles.

The shovel was heated over a fire to dry the sample, which was then tipped into a crucible and heated to red hot, which drove off impurities, such as sulphur and arsenic. It was then treated with a magnet to remove any iron, and what remained would be the valuable tin or copper ore. This was weighed and showed the proportion of the sample that had value, which was then officially recorded. If everything was satisfactory the assayer marked the block with the Arms of the Duchy of Cornwall, using a punch and hammer. This was the official guarantee of both weight and quality. On the last day of the coinage, a crier would be sent around the town to announce its closure. These were busy days, the coinage towns crowded with people. W.G. Maton wrote an account of his visit to the area in his book *Observations of the Western Counties*, written between 1794-6. He was present at one of the midsummer coinages, which were always the busiest of the year, and he described seeing blocks of tin worth 10 to 12 guineas each just lying in heaps all over the town. But, as he pointed out, no sneak thief could get away with a block of tin weighing 320 pounds. One can imagine these days must have been anxious times for the tinners. They had brought in the results of six months of hard work and now they hoped for an appropriate award. There would have been a palpable sense of relief when the weight was agreed and the purity confirmed, and an even greater sense of relief when the sale was completed. One imagines the local inns shared in the delight as their profits soared for those few hectic days of trading.

There was another vital element of the system – the Stannary Courts. These met regularly to hear disputes and they could pass judgement, even if only one of the parties was involved in the industry. In theory, the system was very democratic, and anyone involved as a tinner could bring a case and argue it themselves on payment of one penny. In practice, this did not always work. One can easily imagine a dispute between an ordinary miner and his employer where the one did his best to explain his complaint, while the other hired a glib lawyer to put his case. The Courts did, however, have considerable powers – for example, anyone taking ore from another's bound would be made both to return it and pay a hefty fine. The Courts also had their own gaols at Lostwithiel and Lydford. The Lydford prison had a legendary reputation of being particularly foul and it was said that in the old days, anyone arriving there was hanged first and tried afterwards. If it turned out they were innocent, the authorities paid a priest to pray for them, which apparently made everything all right. The Stannary Courts did not have such draconian powers, but although the gaols were certainly unsavoury and cramped, they were not expected to hold very many prisoners. In fact, many who were sentenced for minor offences were not even locked up at all, but were free to roam around so long as they stayed within the town boundaries. Because courts sat in different places, the gaoler had to go and fetch the accused then from their home towns, for which he was paid tuppence a mile.

The Stannary Courts played a real and important role in the life of the mining community, not least in approving and settling disputes over the bounds that defined the areas in which an individual or group had the exclusive right to prospect and mine. The much grander Convocations of Parliament were established in both counties. Their origins are not really known and claims that they represent an early form of government when the region was independent of the rest of Britain have never been verified. Indeed, the earliest record of the Devon Convocation only dates back to 1497 when the Lord Warden summoned 24 'jurates', representing the different stannary regions, to meet.

Meetings were infrequent, but theoretically both the Cornish and Devon Convocations were entitled to amend or even overthrow laws made by the English parliament, if they affected the tin trade. They were also able to pass laws of their own to regulate the industry. In general, however, meetings were only called when there was legislation to discuss or when the Stannary Courts were considering changes to the system. There might also be disputes about interpretation of the law regarding the courts. The Lord Warden would then decide that a meeting should be held and messages were sent to the Stannaries, informing them they were each required to select six members to attend the parliament, all of whom had to be 'men of property'. There was also a lower house, consisting of six assistant-stannators, from each district, who were all mining experts, and were mainly required to offer practical advice. The Cornish parliament survived right up to the middle of the eighteenth century.

The unique condition of the stannaries and the privileges they gave to tinners, from freedom from certain taxes to a large degree of self-regulation, reinforces the feeling that the tinners were almost a race apart. They had an independence that produced a feeling that spread throughout the community that they had the right to put the interest of the Cornish and Devonian industry before any other considerations – and certainly regarded their own laws as more important than any passed in Westminster. This was to have a real effect on the way in which the industry developed as it grew in size and became more technologically advanced.

Chapter 4

DEEP MINES

The emphasis so far has been almost entirely on tin mining, with scarcely a mention of copper. There was a very good reason for this; copper mining scarcely existed before the beginning of the eighteenth century. A report of 1799 stated that copper was unknown in Britain before the end of the seventeenth century, when it was discovered as a by-product of tin mining, which by contrast had been known 'since time immemorial'. This was obviously not accurate, as a Bronze Age copper mine was discovered at Great Ormes Head in Wales, and in the historic period there was an edict under Henry VIII, banning the export of copper from the country. Carew was writing in 1601:

> 'Copper is found in sundry places, but to what gain to the searchers I have not been curious to inquire, nor they hasty to reveal; for of one mine of which I took view, the ore was shipped to be refined in Wales, either to save cost in the fuel, or to conceal the profit.'

The strong probability was that refining was taking place in Wales, because it was cheaper to send ore there than it was to import coal. Unravelling the facts is quite complicated. Another writer, William Borlase, an amateur geologist and Rector of Ludgvan, wrote in 1758 that some forty years earlier 'a certain Mr. Costar' had invented a 'new water engine' that had allowed him to drain a mine, and it was from that date that the price of copper went up and mine improvement began. 'Water engine' was often used as a description of some form of steam engine, but there appear to be no records of any form of steam engine being used at this early date.

The general consensus seems to be that, as a report to a Committee of the House of Commons stated, that copper appeared first in mines being worked originally just for tin. A charter was granted as early as 1595 for manufacturing brass, which is an alloy of copper and zinc, but the first charter for refining copper ore in Britain only dates to 1691. In spite of the claims for the Costar draining engine, it is likely that the actual development of copper mines depended mostly on the development of an efficient draining system for deep mining. And finding that was to prove the starting point for a revolution in the world of power.

The simplest form of pumping system was the rag and chain pump. This consisted of a continuous loop of rope or chain with rags tied on in tight knots at various intervals. The loop passed round a wheel, and through a flanged tube, the lower end of which was below the water. As the loop was turned by the wheel, so the rags pushed water up the tube to be discharged at the top. A more sophisticated version was the bucket pump, which plunged down into the water, which entered through a valve that then closed, allowing the water to be raised on the upstroke. The system used in mines in the seventeenth century often involved bob pumps, worked by water wheels. The system is basically simple. You start with an overhead beam, pivoted at its centre, much like a giant seesaw. The pump rods going down the shaft are hung from one end. Their weight will naturally pull the beam down at that end. What is needed next is some form of power to pull down the opposite end of the beam to lift the rods to make the pumping system work. A simple crank worked by a water wheel could provide that movement. Water power worked up to a point, but as mines became ever deeper was less effective. Something new was needed.

The first attempt to produce something entirely new was the work of a former military engineer, Thomas Savery. In 1698, he patented his 'Miners Friend and Engine to raise water by fire'. The system consisted of a large boiler set at the bottom of the shaft. Steam was forced up into a higher vessel, where it was condensed, creating a partial vacuum. Air pressure then forced water up a pipe into the vessel, until it was full. The process could then be repeated, forcing the water ever higher. It was not very satisfactory, partly because it was very wasteful in terms of fuel use and also because the technology for containing high pressure steam was scarcely adequate. It was never widely adopted. Soon afterwards, another inventor turned instead to the simpler idea of finding a way to work the familiar beam pump.

Thomas Newcomen was born at Dartmouth and baptised in 1664. He is said to have had an engineering apprenticeship in Devon, but his name is not mentioned in the records. We do know that he had an ironmongery business in Dartmouth, from which he supplied tools to the mines of Devon and Cornwall. He was thus well aware of the problems of mine drainage. He set about devising his engine. This was to work the type of beam or bob pump described above. A piston, fitting snugly into an open topped cylinder, was suspended by chains from the opposite end of the beam from the pump rods. Steam at low pressure was raised in a simple boiler and fed into the cylinder below the piston. It was then sprayed with cold water and, as in the Savery engine, the steam condensed and a partial vacuum was created. Air pressure then forced down the piston, raising the pump rods. Pressure equalised, gravity took over again and the rods dropped – the cycle was complete. The machine could be made self-acting by means of rods from the beam that could open and close valves. The first engine was built at a colliery at Tipton, close to Dudley Castle, so is usually known as the Dudley engine. The cylinder was 21 inches in diameter and 7 feet 10 inches high. It worked at 12 strokes a minute and could raise 10 gallons of water to a height of 153 feet at every stroke. A replica of the engine can be seen at the Black Country Museum. An original Newcomen engine of c.1725 can be seen in a special museum at Dartmouth. Because the actual work is done by air pressure not steam pressure, the Newcomen engines are usually called atmospheric engines.

The Newcomen engine was an undoubted success, and the Godolphin family, who had an estate near Helston, had an engine erected at Wheal Vor as early as 1715. It drained water from a depth of 60 fathoms below adit. But at first it seems only a few such engines were built in Cornwall, though the exact number is uncertain. The problem was that they consumed vast quantities of fuel, and although there may have been attempts to use turf as a fuel, in general coal was required – and that had to be imported from South Wales. To make matters worse, there was a tax on Welsh coal, apart from use in smelting. In 1739, the Cornish successfully lobbied the government to have the tax removed. It was good timing, for the price of copper had soared and there was a rush of engine building that enabled new mines to be opened up and old mines that had been considered unworkable to be drained and started back in production. This provided openings for engine construction in Cornwall, and a number of new concerns were established. John Wise came from the Midlands to set up in business in Truro and two other manufacturers, John Budge and John Nancarrow, also seem to have come to Cornwall from the Midlands, though little is known of them. The most important manufacturer was Joseph Hornblower, who was originally from Broseley in Shropshire and came to Cornwall in the 1720s to erect engines before going into manufacture. By this time, construction had improved. Where early engines had to be cast using brass for the cylinders, they could now be cast in iron, thanks to the new coke fired blast furnaces developed first by Abraham Darby in Coalbrookdale. Joseph's son Jonathan continued the business – and is also well known for a curious quirk. He had thirteen children, who all had names beginning with J.

The Newcomen engines did their work and where extra power was needed, they were simply made with larger cylinders. The largest ever recorded was erected at the Chacewater Mine in 1775. Built by the Carron Iron Company in Scotland, it had an 82 inch diameter cylinder, a huge increase over the 21 inches of the original Dudley engine. But there was a price to pay; greater size meant increased fuel costs and there was an incentive to try and make the engines more efficient. The most successful engineer in this field – as he was in other branches of engineering – was John Smeaton, who designed the Chacewater engine. He had earlier carried out experiments on wind and water mills, during which he had dispelled the accepted views by proving that overshot wheels were the most efficient. Now he turned his attention to the atmospheric engine. He managed to improve the basic atmospheric engine performance by around 75 per cent. He also came up with a measure of power, based on the work a horse could do in an 8-hour day which he estimated at 22,000 foot pounds per minute. James Watt would later revise this measure to 33,000 ft lb/minute, the measure still in use today for horse power. Smeaton's Chacewater engine was estimated at 72 horse power, roughly 53 kilowatts in today's units. With massive engines to do the pumping, there seemed to be no limit to the depth to which mines could be sunk – provided the money was available for development. The mining world of the South West was changing.

Just as mine drainage was changing, so too were working methods. In the early days, when men were met by hard rock, they had the greatest difficulty in dealing with it. One of the oldest methods, certainly used by the Romans in their mines in Britain such as the Dolaucothi gold mine in Wales, was fire setting. This simply involved just what the name suggests – lighting a fire beside the rock face and then when the stone was red hot dashing water against it to create cracks. There is not much evidence of this being widely used in tin and copper mines. The most common method was exactly the same as that used in stone quarries. A hole was drilled by hand with a tool that had a cruciform bit. Into this was inserted feathers, semi-circular iron rods. A steel wedge was then set at the gap between the feathers and hammered home, driving the feathers apart to split the rock. It was slow and laborious work.

At the very end of the seventeenth century, German miners brought a new technology to Britain – blasting with gunpowder. Prince Rupert of Bohemia, who had financial interests in mines in both Germany and England, introduced the material to this country. The Godolphin family, who had first introduced the Newcomen engine to Cornwall, were also the first to introduce gunpowder into the county. They arranged for an expert from the Mines Royal of Somerset to come to Wheal Vor to instruct the local miners in 1689. Thomas Epsley is also credited with introducing black powder about 1695, but he died in a mine accident shortly afterwards. Gunpowder was also definitely recorded as being used at a mine near St. Agnes. It did not remove all the hard work. The miners still had to drill holes by hand, a particularly difficult task when drilling upwards, and the hammer had to be swung underhand, but it did have the advantage that the hole was self-cleaning, the debris simply falling out. One man would hold the steel against the rock, while the other drove it home with a six-pound sledge hammer. If the rock was wet, clay had to be rammed in with a claying bar, so that it filled up any cracks and kept the hole completely dry. The hole would then be packed with the black powder, and a simple fuse added, made from quills and filled with more powder. A narrow rod was pushed through the powder and carefully removed, leaving a hole into which the fuse could be slipped. Then the man who had merely held the steel got the job of firing and clearing away the broken rock. The dangers associated with blasting are all too obvious. A candle was lit next to the fuse, and the flame only reached the powder once it had burned down a fixed amount. This, in theory, gave the man lighting the fuse enough time to scurry to safety as quickly as possible. Theory and practice did not always coincide. There were less obvious perils. Often the powder was rammed home with an iron rod and if that should happen to strike the rock there would be a spark that could set off the explosion. In a report from a coal mine, not a metal mine, there is a gruesome account of just such an event, when the rod was shot straight through the unfortunate miner's head.

William Pryce described the system for dealing with the various types of accidents as it existed at the latter part of the eighteenth century. In the big mines, employers levied 2d a week for the use of a surgeon to attend all accidents. But in Pryce's time it seemed that surgeons were reluctant to take on the work, though their services were clearly essential.

> 'Now, in the course of a year, it is three hundred to one, that the trepan, or the crooked knife, will be wanted not only once or twice, but very often; besides the ordinary accidents of burns, wounds, contusions, laxations, or simple and compounded fractures, where the knife is spared; and the blasting one or both eyes, and the two last fingers of the left hand, by gunpowder.'

Pryce also bemoaned the fact that when a miner did have an accident, the treatment was not the end of his troubles. 'The patient is then conveyed six or seven miles to his own hut, full of naked children, but destitute of all convenience, and almost of all necessities.' His solution would be to establish a county hospital and, as he pointed out, given the wealth created by the industry it should be affordable. He does not seem to have been equally concerned by the home conditions he described. Later he went on to describe fevers, which proved fatal in many cases, and which appeared to come in epidemics that spread through the community. Pryce put this down to some kind of 'effluvium' emanating from the mines, but was far more likely to be caused by such factors as contaminated water in the miners' houses. But even without being struck by diseases or accident, the miners' lives were comparatively short, as one local reported a man of fifty was regarded as 'very old'. And that was always attributed to the working conditions. A traveller in the 1790s, E.D. Clarke, in his book *A Tour through the South of England*, published in 1793, wrote:

> 'At about eighty fathoms depth we came to a vein of copper ore where two sorry wretches were busied in the process of their miserable employment. With hardly room to move their bodies, in sulphurous air, wet to the skin, and buried in the solid rock, these poor devils live and work for a pittance barely sufficient to keep them alive; picking out the hard ore by the glimmering of a small candle whose scattered rays will hardly penetrate the thick darkness of the place.'

The working of deep mines only served to make conditions worse and the work harder than it had been in earlier times. Although Pryce has little to say about the working conditions, he tells us quite a lot about how the mines developed, starting with the sinking of shafts.

> 'Those shafts in deep mines, are often costly and troublesome to be sunk from the surface of the earth; either by means of the water that falls into them, the intense hardness of the stratum they must cut through in sinking, or by means of loose soft ground that requires much timber and boards to line the shaft from top to bottom, When they find any of these difficulties very great they sink a little-winds in this manner they go down in the grass shaft, from whence the lode is gone as far as the shaft is perpendicular, or as far as they think proper; from these they work in a drift or horizontal line,

> till they come as far over the underlie of the lode as they like; there they cut a plot and in the middle of this plot they fix a windlass on the winding tackle and sink down their little winds or shaft until they cut the lode in it, or to the depth they intended.'

Basically, Pryce is describing a system that created a three-dimensional underground labyrinth. Because the veins of ore were so often found to be sloping quite steeply, it was necessary to cut horizontal passages from the main shaft, and from these drop subsidiary shafts to meet the lode – the winds or winzes. The plots he mentions were an essential part of the system. A huge amount of waste rock had to be removed to get to the ore and it was too expensive and time consuming to bring it to the surface, given that the only mechanisms available were simple windlasses or horse gins. So the miners created a cavity in the rock, generally about twelve feet square and six feet high, in which the waste could be stored. The same name was used for the excavated areas from which the subsidiary shafts were sunk. After the ore had been removed, some rock would be left standing, like 'the pillars of a venerable piece of Gothic architecture' to prevent the roof falling in. If this was insufficient, then timber props would need to be added.

One of the problems encountered as mines grew ever larger and employed more and more men, was getting men and material up and down the mine. Originally, everything would have had to use the same main shaft, but as the mine system became more complex, it was possible to utilise an old shaft and use it exclusively for getting the miners in and out of the workings. The system that developed was to build a number of wooden stages, each connected by ladders, from the surface to the lowest level of the mine. It was a logical system, but one that caused great hardship for the workers. William Crago, looking back on his childhood when as a nine year old boy he joined his father at work, gave a graphic description of what it was like using the ladders. His father had a load of heavy equipment to carry, but the young boy was also expected to do his bit, even though it was his first time down the shaft. First, he collected candles that were strung round his neck, then he was loaded up:

> 'I had on my right arm about five pounds of black powder carried in a copper can, on my left, a coil of fuse to be used with the powder for blasting purposes, in my pockets gad, used for splitting rocks and each of us a fair sized potato pasty for our dinners.'

His father waited until all the other men had started before setting off, so that there would be no one behind them harrying the young lad to hurry up. After descending twenty feet, they paused to light the candles stuck on their hats by a lump of clay:

> 'We at last stepped into the footway, Father first and I following him, very carefully we descended the first 480 feet. It was almost like climbing down the side of a house and as we slowly went along, ever and anon came Father's warning voice 'Hold tight your hand, my son'.
>
> 'As I have before said, the first portion of our downward course was like climbing down the side of a house. The portion of the shaft we were now about to descend was termed the underlie Section … When the ladders are upright the strain of climbing comes almost entirely on one's arms whereas if the ladders are on the slope the feet have to bear a much larger portion of the weight of the climber's body, and so I found on resuming my descent, it was very much easier to climb down this portion than had hither found it, but with that advantage I found there was attached to it a great disadvantage. You will readily imagine that in most mines there is a great quantity of water and as a rule Cornish mines are very wet indeed.
>
> 'It is not only the water that is found while sinking the shafts and driving the levels, but surface water which finds its way down, gives a great deal of trouble to the miner engaged in sinking operations and very much inconvenience when climbing up or down.
>
> 'It is not pleasant to have a small stream of water dropping just on the back of your neck and running down your back and legs into your boots and then have to work in that state for 7 or 8 hours, but such was the fate of most of those who worked in that portion of our mine.
>
> 'We continued what seemed to me our interminable journey down into blackness, my legs were aching, my back was stiff and sore, and my hands were getting benumbed with the continual clutching of the ladder rungs and it was quite necessary that Father's warning shouts be given "Hold tight your hands".
>
> 'After about two hours climbing (I was soon able to do it as quickly as most boys) we reached 1,600 feet landing stage and I for one felt extremely thankful to step out of the ladder onto what appeared a much safer place, but in order to enter the subway we had to cross the shaft on what appeared to be a very narrow plank and great caution is necessary in so doing, however, this was safely accomplished and we entered the mouth of the subway.'

The boy now had to work his long shift underground, and he then had the arduous task of climbing all the way back up again – and that at an age when a modern child would still be in primary school.

Working conditions in the mines in Cornwall and Devon were bad but were not necessarily worse than those in any other mines, and in some respects far better than in collieries. They did not face the risk of explosions from what was known as 'fire damp', methane gas escaping from the coal seams, or 'choke damp', carbon dioxide that could suffocate. They were not immune from the miners' great curse, silicosis that came from the clouds of tiny stone particles that filled the air during blasting and drilling and ended up in the workers' lungs. But they did enjoy a measure of control over their own working lives that was denied to the vast majority of other miners.

Chapter 5

TUTWORK, TRIBUTE AND DRESSING

The system of organising the work of the mines that was unique to the region developed over a period of time. A French author, Gabriel Jars, described the system as he found it in the mid-eighteenth century in his book *Voyages Métallique.* This was the version in which the adventurers, in effect the men who owned the mine on lease from the landowner, held a Dutch auction, at which the contractors put in their bid to work a pitch, driving a level for an agreed length, varying from fifteen to twenty-four fathoms. They would then employ men at an agreed wage to do the work. This is the system that had been in place for some time, but Jars also mentioned that sometimes the contractors were not individuals but a consortium of workers who were not on wages but shared the profits from the work. Making a bid was a gamble. They could draw on their own experiences to guide them, but there was no way of knowing what difficulties might lie in their way once they started work. This balancing act between needing to make a low enough bid to secure the contract, but not so low as to risk making no profit or even a loss was at the heart of the system that developed in the latter part of the eighteenth century.

John Taylor (1779-1863) was a notable figure in the history of tin and copper mining. He developed a machine for crushing copper ore, known as the 'Cornish rolls', in 1796 and in 1798, when just 19 years old, he was appointed as manager of Wheal Friendship near Tavistock. He was also the engineer responsible for the construction of the Tavistock Canal that joined that town to Morwellham Quay. It not only served its purpose in transporting ore, but during its construction, valuable lodes were discovered. He was thus well qualified to provide a guide to the system of payment used in the mines, which he did in a long article in the Transactions of the Geological Society, 1814. He began by stressing the unique values of the system as he saw them.

> 'We now come to that part of the economy of the Cornish mines, which is most deserving of consideration from the effects it has produced, not only by procuring regularly a great deal of effective labour in proportion to the money paid for it, but also by turning that labour into such a direction as

> to make it the interest of the workmen to increase the discoveries of ore, and to work it and make it saleable in the most economical manner. Thus the owners of the mine have the advantage of all the intellect and skill that the men collectively possess, and have only to guard against the chances of fraud which such a system may be supposed to be subject to, but which are in fact under intelligent and faithful agents of too trifling a nature to be accounted of any importance.'

It is interesting to note the change in tone from Pryce's descriptions of workers and mine captains. Taylor writes of the intelligence and skill of the workforce and the capabilities of mine captains, where Pryce tended to denigrate the miners and regard all captains as potential rogues and thieves. But by this time, systems had changed. Work was still let by public auction, but was divided into three distinct categories as Taylor succinctly explains:

> '*Tutwork* includes work done by measure, such as sinking shafts, driving levels, or stoping ground; the first being paid for by the fathom in depth, the second by the fathom in length, and the third by the cubic or solid fathom.
>
> '*Tribute* is payment for raising or dressing ore by a certain part of its real value when merchantable, and it is this part of the system that is deserving of the most attention, both on account of the excitement it produces to discover ore and to raise it cheaply, as already noticed, and on account of the perfect state to which the arrangements with the working miners under this head have been brought.
>
> '*Dressing* contracted for at the surveys is seldom for more than the waste or ore leavings of the tributor, the ores raised on tribute being made merchantable under the same contract; but as the men working on the terms usually made, cannot often afford to dress the coarser parts of what they raise, reject it, and it is left to others who stamp and clean it, having a proportional price likewise in the way of tribute.'

Tutwork was divided into 'bargains', a specific job requiring a set number of workers always referred to as a pair or pare, no matter how many may actually be used. In sinking shafts or driving levels, the numbers are limited by the actual working space. So shafts may require anything from four to twelve men and levels two to six. Stoping involved the actual business of removing the ore from the lode. One of the pair had to act for all of them and was known as the taker. Tribute work was set in pitches, clearly defined areas of the mine and usually worked by gangs of two to six men. Dressing is, like tutwork, set in bargains and went generally just to one man, who then employed women and children to do the actual work. This is one of the earliest references to the work of women in the industry in treating the ore. They were known as 'bal maidens' – bal being a Cornish name for a mine and maidens because they usually gave up work when they got married. By 1800, it has been estimated that there were some 2,000 girls and women employed in ore dressing in Cornwall. We shall be looking at the actual work they did later in the chapter.

The system was quite complex and began a day or two before the actual bargaining got under way. The mine captain would have to carefully investigate what needed to be done by both tut and tribute and form his own estimate of how many men would be needed and what would be a reasonable price to pay for the work. He also had to take responsibility for measuring up the tutwork carried out in the previous period, so that payments could be calculated. All this involved a lot of complex calculations, and the count house was an important part of any mine. On the day itself, the men gathered together and the process began with reading out the rules and regulations, which was something of a formality, as the conditions rarely varied. Then the tutwork was offered up. The first work on offer might be for continuing work on a shaft or drift that had already been started. Usually the men who had already been at work started the bidding at a very high figure, which they were well aware they would not get. This was a simple ploy. If they started at a modest price, there would be little room left for manoeuvre. Bidding went on until a bottom price had been reached, and no bidder was prepared to go any lower. At this point, the captain threw up a stone to indicate the bidding was over and the representative of that group had the option of taking on the work. It did not, however, mean that the adventurers were forced to accept the bid. They would have had their own estimate of what should be paid, so there was more toing and froing over the process until agreement was reached. A similar process was followed when the tribute work came up.

It might seem that once agreement had been reached everything was very straightforward, but the successful bidders were far from certain how much money they would actually make at the end of the period, usually two months, for which the contract had been agreed. The men could not be sure whether the material through which they had to hack their way would be easy to work, loose material or hard rock. Depending on how things worked out, they could either make a good distance through easy ground and make a decent amount of money, or be severely hampered and get far less for what would have been far harder work. Regardless of how the work turned out, they had to pay money to the company for the essential tools and equipment they needed. At the end of the period, a 'bal bill' was drawn up setting out all the details. The following example is for a four man team working at Carn Brea in 1863 for May and June.

£ s d
One fathom 0 feet 7 inches at £12 a fathom 13 3 4
One fathom 9 feet 0 inches at £13 a fathom 13 0 0
Assisting shafts men 1 15 0

---

27 18 4
Sundry deductions:-
£ s d
To candles 1 19 4
To powder 1 13 4
To safety-fuse 0 6 0
To sundries 0 1 0
To smith costs 1 5 5
To drawing 0 6 3
To tramming 0 0 0
To doctor and club 0 6 0

---

5 17 4

The men thus received a total of £22 1s to share between themselves for their two months' work, which at today's prices would work out at roughly £80 a week each. That is not quite the end of the story. During their working period no money was coming in. They usually had to borrow money, known as 'subsist', from the company that had to be paid back with interest, effectively reducing their income still further. The bill quoted here is for a group of four men, but it was more usual to have six, so that the work could be divided into three 8-hour shifts for every working day of the contract.

Tribute work was far more complex, as the money the miners received depended not just on the amount of work they put in, but also on the quality and quantity of the ore they recovered and the price paid when it was sold. So, anyone bidding for a pitch had to make an estimate of a whole range of different factors. Would the ground be difficult or easy to work? Was the lode likely to remain unchanged as they advanced and if it changed would it get better or worse? How much would it cost them to get the ore to the surface and have it dressed? What was the market price of metal doing – rising or falling? If they got their calculations wrong, then they could end up working for a pittance or even losing money. On the other hand, if they had been astute or lucky, the quality of the ore could be better than expected, the work easier and the market price could rise. In a sense, it was a lottery, but it was a lottery in which the skill and experience of the miners played a crucial role.

John Taylor explained the obvious advantages in the system as he experienced it in the copper mines. The miners are encouraged to seek out the best ores and work them as efficiently as possible, as their payments are decided as a previously agreed proportion of the price when the ore was sold. Everything had to be accurately measured. In the copper mines, the ore had not only to be mined, but brought to the surface and dressed ready for sale. Each individual parcel was brought to a mine captain, who weighed it, taking a sample which went for assaying to determine its value, and was then added to the general store, the 'public parcel'. No payment was made to the miners until the ore was sold to the smelter. As with tut payment, there were deductions to be made first and in this case, the costs were considerably higher. The men had to pay for tools and equipment as with tutwork, but they also had to bear the cost of dressing the ore themselves. Similar bal bills to that quoted above for tut work were drawn up for tribute. Two men working at Basset Mine in the 1860s brought up ore that sold for £20 18s 6d and for which their share was seven shillings in the pound - £7 6s 11d, but although the work was completed at the end of October, payment was only made on 27 November. Apart from the usual deduction, they were also charged 12 shillings for drawing the ore up to the surface and another 12 shillings for dressing. By the end of that process, their earnings had shrunk to £5 3s 9d.

Taylor points out that there have been occasions where men, like those quoted above, had agreed on a share of 7 shillings in the pound, but then discovered a very rich lode, which if the owners had known its value, would only have paid 2 shillings. On these occasions, it is not unknown for men to earn a hundred pounds each in a week. But, as he points out, it is far more common to find a lode that failed to meet expectations, in which case the men not only received a very small amount, but it may not even be enough to cover the expenses they had accrued. In other words, they had no money and debts still unpaid that would accumulate interest.

Taylor expressed his view that, on the whole, the system favoured the adventurers, even when the miners got lucky – or as the men might put it, showed their shrewdness and good judgement. The big payments were rare and when they did occur, they benefitted the owners, who after the initial payments could work the lode at their own prices. And it was rather like a casino when someone hits the jackpot; everyone else is encouraged to spend more in the hope of being the next lucky winner. In this case, all the men worked that bit harder, hoping to find their own fortunes. There was another, darker, side to the system. Many of the employers were also the suppliers of essentials that the miner needed for work and subsistence, so that in effect they had a monopoly and could charge more or less what they liked, supplying goods on credit and collecting interest on the debts that were built up. It was not uncommon for a mining family to end a stint at work and, having paid off what they owed for the previous two months, have nothing left and no option but to fall back into debt. A bal bill for Dolcoath in 1842 records that four men only earned £4 18s 9d but their expenses amounted to £4 14s 6d and on top of that they had needed to borrow a 'subsist' of £7 6s 6d. All their efforts had left them over seven pounds worse off than if they had sat at home doing nothing.

One of the features that many commented on was the lack of strikes and trade unions in the industry. This is not too surprising, as the workmen instead of co-operating were divided up into small units competing against each other. Workers did not receive a straight wage but instead had payments related directly to their own efforts and the value of what they produced. They were, as one commentator put it, 'their own employers'. The other feature that comes up time and time again is the independence of the workers, who were never simply following instructions but were constantly having to make their own judgements. As one writer put it:

> 'It may fairly be asserted, that the solution of every intricate problem in mining geology is generally assigned to a Cornish agent, and every task requiring skill, resource, and courage intrusted to a Cornish miner.'

Ore dressing was also carried out by tribute in most cases, but here it was generally a comparatively wealthy employer who took on the work and paid fixed wages to the workers. The majority were the bal maidens, girls who often started work at ten years old, though sometimes even younger. The ore would have to be broken into manageable pieces at the mine for transport, and when it reached the dressings would be brought to the workplace in barrows that had two pairs of handles but no wheels. They were carried by older, stronger girls who worked in pairs carrying a hundredweight and a half in every load. At tin mines, the main work was 'spalling', breaking down the ore with long handled hammers, after which it could be taken to the powered stamps. In copper mines, the system was usually different. The ore went first to young girls to be sorted in water, an unpleasant and dirty task at the best of times but in winter with the water freezing cold must have been far worse. It was skilled work but poorly paid, even by the low standards that applied to all the womens' work. Then the process of breaking down the ore went through two stages; first the 'cobbers' used short handled hammers to break the ore into nut-sized pieces, after which it went to the 'buckers', who reduced it to a powder with flat-headed hammers.

The final stage in processing the ore came when the mines sold it on to the smelters. Here there was no distinctive work system, simply owners taking the profits and workers on fixed wages. In the eighteenth century, all tin was smelted in Cornwall, but almost all the copper ore was sent to South Wales, mainly to Swansea, because the process used prodigious quantities of coal, readily available from the local Welsh pits.

The eighteenth century had seen the industry changed radically and its unique mixture of tut and tribute work established. From around 1750 to the end of the century, the production of Cornish tin remained more or less static, with 2,786 tons produced in 1750 and after a brief period in the 1790s where production rose above 3,000 and in one year reached a height of 3,809 tons, by the end of the century it had dropped back to under 3,000 tons. The picture in the copper industry was very different. From a negligible amount at the beginning of the century, by 1800 production had risen to 55,981 tons of ore yielding 5,187 tons of copper. There was also to be another vital change in the industry towards the end of the century, when the atmospheric engines began to give way to the genuine steam engine.

Chapter 6

WATT AND THE PIRATES

For most of the eighteenth century, the use of steam power for pumping had been limited by the huge expense in running the atmospheric engines. In spite of the improvements made by Smeaton, the engines could not work to a depth below adit of around 600 feet – which effectively prevented working any lodes deeper than that. It was a useful but inefficient machine – in fact, in its original form, it had a thermal efficiency of just 1 per cent. In other words, of the energy produced by burning coal to raise steam, just one hundredth was translated into useful work. There was a fundamental flaw, and it was a young instrument maker at Glasgow University who not only identified the problem but found a solution. His name was James Watt.

During the academic year 1763-4 he was sent a model Newcomen engine that was simply not working as it should have been. Among those who Watt met at the university was Professor Joseph Black, who had recently described the phenomenon that he called 'latent heat'. He had discovered that extra heat was needed to turn a material from one state to another, such as water into steam. Whether Watt had known of this work or not, latent heat was the problem. Extra heat was needed in the Newcomen engine because at every stroke, the cylinder was cooled down and had to be reheated. He pondered over the problem, and then one day the answer came to him, as he later described:

> 'It was on the Green of Glasgow I had gone to take a walk on a fine Summer afternoon, early in 1765. I had entered the green by the gate at the foot of Little Charlotte Street and had passed the old washing-house. I was thinking upon the engine at the time and had gone as far as the herds' house, when the idea came to my mind that as steam was an elastic body it would rush into a vacuum, and if a communication was made between the cylinder and an exhausted vessel it would rush into it, and might there be condensed without cooling the cylinder.'

He had come up with a completely new idea – the separate condenser. By condensing the steam outside the cylinder, the latter could be jacketed and kept permanently hot. There was, however, still a lingering problem – heat would inevitably be lost through the open top of the cylinder. But if he closed the top, then air pressure could no longer be used to push down the piston. He then had the revolutionary idea. He would dispense with the need to use air pressure at all. Instead, he would feed steam under pressure into the space above the cylinder where it would expand and provide the necessary force. He still had to make his new idea work in practice and he had to invent another new device, somewhat confusingly known as the air pump. It was not needed to pump air, but to remove condensed water.

There is a saying that if you invent a better mousetrap, the world will rush to your door to buy it. There was no stampede to Watt's door. He had hoped to develop his engine with the help of Dr John Roebuck but seemed to be making little progress and was forced to take on other work to make a living, including a stint as engineer in the construction of the Forth and Clyde Canal. It was while going to London to give evidence on the Canal Bill to a Parliamentary Committee that he stopped off at Birmingham and visited the Soho factory of Matthew Boulton. There Boulton manufactured 'toys' – not playthings for children but small items such as gilded buckles and buttons. He wasn't there in person but Watt had a long talk with Boulton's friend and associate, Dr William Small. Obviously, the engine was discussed and Small promptly wrote suggesting a partnership. Boulton and Small were offering what Watt sadly lacked – proper funding. 'In a partnership I liked,' wrote Small, 'I should not hesitate to employ any sum of money I can command.' Nothing came of this immediately, but soon Watt was writing to Boulton suggesting a partnership that would manufacture his steam engines for the Midland counties. That was not at all what the ambitious Boulton had in mind:

> 'My idea was to settle a Manufactory near to my own by the side of our canal [the Birmingham Canal] where I would erect all the conveniences necessary for the completion of the engines and from which Manufactory we would serve all the world with engines of all sizes'.

It was an offer far too good to resist and soon Watt was moving from Scotland down to the English Midlands to take up his role in the partnership of Boulton and Watt, manufacturers of steam engines. James Watt had taken out a patent in 1769 for his invention that was so sweeping in its details that it made it impossible for anyone to build any engine featuring the condenser or air pump and prohibited any attempts to change or improve on Watt's own design. This was to prove highly controversial and resulted in what amounted to a battle of the engineers in Cornwall.

Watt might have been the inventive genius, but he was no businessman and it was Boulton who realised that one of their most profitable areas for sales would be the mining region of the South West. He set up a system by which anyone wanting an engine had to do so on the company's terms. They had already discovered that the Shropshire iron master John Wilkinson was the one man capable of providing accurately bored cylinders and all orders had to go to him. The precision engineering parts, such as valves, were manufactured in the new Boulton and Watt factory and again no other suppliers could be used. There were other parts that could be made on site by whoever bought the engine, but they gave precise instructions on how everything was to be done, from manufacturing the beam of the beam engine out of a single piece of seasoned oak to constructing the engine house itself. All the work had to be carried out under the supervision of a competent agent appointed by Boulton and Watt.

A crucial part of the agreement concerned the payment by the mine owners to the Birmingham partners. This was a royalty known as the premium, based on a proportion of the savings to be made by changing from a Newcomen atmospheric engine to the new steam engine. The efficiency was measured in terms of duty. This was defined as the number of pounds of water that could be raised one foot by the consumption of a bushel of coal. The early Newcomen engines had a duty of roughly 4.5 million. This was improved by Smeaton to 12.5 million, but when the first Boulton and Watt engine was installed at Bloomfield Colliery at Tipton in the Black Country it was rated at 22 million and that figure was soon improved. This was the sort of statistics that had engineers in Cornwall sitting up and taking notice. The agent for Wheal Virgin wrote in despair, 'The existing engine at full power and the mine will be forced to stop unless some answer is found. The engines last month used 300 Weys [30 tons] of coal … which sweeps away all the profits.' He wondered if it was possible to stick one of these new condensers on the side of his old engine and save himself the expense of buying a new one – an idea that, even if it had been practicable, was hardly likely to be recommended by Boulton. Not all, however, were immediately convinced that the Boulton and Watt engine had any merit at all. One engineer wrote in to say that he got just as good a vacuum using his old engine, proving that he had completely misunderstood what Watt had achieved.

The first Boulton and Watt engine was installed at Wheal Busy at Chacewater near Redruth in 1777. Very soon, news of the savings being made and the increased efficiency of the engine became known throughout the mining districts and orders began to flow. Over the next quarter of a century, some fifty engines were installed in Cornwall, with cylinder sizes varying from a mere 12 inch to a massive 64 inch diameter. There was initial enthusiasm and it seems the miners of Cornwall were more au fait with the new technology than the men who worked at the Hawkesbury Colliery near Coventry, where an early engine had been installed. James Keir, a business associate of Boulton's, wrote to him:

> 'It gave great pleasure to hear of the propriety of affairs in Cornwall & of the reasonable disposition of the Inhabitants. It must make Mr. Watt's residence much less disagreeable than if he had found them such a medley of Fools & Knaves as they are at Hawksbury.'

Cornwall was such an important customer for engines that it was decided the company needed a permanent agent in the region, both to look for new orders and to supervise the erection and running of the machines. The first to be appointed was a local man, Thomas Wilson, who was later joined by the Scotsman William Murdoch, who later began spelling his name 'Murdock'. The story of how Murdoch came to be employed is not strictly relevant to this narrative, but is too good to leave out – and Murdoch was to play an important role in the developing relationship between the Soho company and the mining engineers of Cornwall. It seems Murdoch had heard about what was being done in Birmingham and decided to walk from Scotland to try and get a post there. He needed something different to attract Boulton's attention, so he appeared for an interview wearing a wooden hat. This intrigued the entrepreneur and when he heard that Murdoch had made it himself, gave him a job. The tale may be a myth but Boulton quickly recognised that this was, in his own words, 'a valuable man'.

Although they had trusted representatives in Cornwall, both Boulton and Watt thought it essential to make periodic visits to the region themselves. They had difficulty finding good accommodation but Watt, who had remarried in 1775 following the death of his first wife, persuaded his wife Ann to accompany him. It was not a success. She admired the coastal scenery, but they were forced to spend most of their time in the heart of the mining district. Nothing pleased them, and they were scathing about the servants they employed, 'a set of Laziest Wretches that now Breathes on the Earth to say nothing of their Nastiness'. Matters were improved when they were able to take a lease on a comfortable house near Truro. There is an interesting side light to these visits. Watt needed to keep up his business correspondence, but no longer had a clerk to produce copies. So he invented a copying machine, which he later patented. It went into production and the company sold 630 machines in the first year at seven guineas each. They remained in use until the invention of the typewriter and carbon paper; a profitable little enterprise.

The first Boulton and Watt engines were, like the Newcomen, single acting; that is the power was only applied in one direction. In order to work it still needed the weight of pump rods on one end of the beam. The reason is obvious. The piston was attached to the overhead beam by a chain – you can pull down with a chain but you can't push up. The answer might seem to be to simply replace the chain with a rigid metal rod, but then another problem appears. The piston moves up and down vertically, but the end of the beam moves along a curved path. Watt solved this problem in 1784 with what he referred to as his finest invention – the parallel motion. This was quite a complex system, with a moveable parallelogram of rods, with the connecting rod from the piston attached to one corner. The linkage shifted as the beam moved, keeping the connecting rod moving almost in a straight line, and near enough vertically so as not to put pressure on the piston rod.

Now steam pressure could be used on either side of the piston to drive it up and down; the engine was double acting. This meant that the opposite end of the beam, instead of being only attached to pump rods, could now be supplied with a simple device such as a crank to create circular movement. The steam engine could now take on any job that would previously have relied on animal or water power. It could be used to haul material up and down a shaft as a whim engine or to power stamps. A small whim engine – not Boulton and Watt but similar – has been restored and is regularly steamed at the Levant Mine on the north coast near St. Just. No stamp engines have survived, but there are sites with all three engine houses, pumping, winding and stamps, still standing, such as Wheal Peevor.

Everything seemed to be going well with both tin and copper mining until 1778, when Thomas Williams, an Anglesey solicitor, formed the Parys Mine Company to extract the ore from Parys Mountain on the island. It was not strictly speaking a mountain at all, as it only reached a modest 500 feet above sea level, but it was rich with ore. Soon men were at work quarrying, digging shafts and tunnels. Great chunks of the hill were removed and visiting the site today is like coming across a lunar landscape from which all soil and foliage has long since been stripped. At the height of activity there were over 1,500 men at work and the mine was producing a thousand tons of ore a year and because it was all comparatively near the surface, it was cheaper to work than the deep Cornish and Devon mines. Even transport was easier; a little harbour was established at Amlwch near the foot of the mountain for shipping the ore. James Watt was one of those who saw clearly what it would mean for Cornwall. The rate at which ore was coming from Anglesey brought down the price of copper. He declared that the situation 'makes one tremble for the business as a whole'. The Cornish tried to compensate by increasing tin production, which only had the effect of lowering that price as well. The whole region was plunged into an economic crisis, as explained by Nicholas Donnithorne, chairman of the Quarterly Tin Meetings in Cornwall, who wrote in 1789:

> 'I am acting on behalf of many thousands who at this time are destitute of even most common necessaries of life. I am lately returned from the mining parishes in Cornwall, where I have ben witness to the greatest imaginable poverty and distress, insomuch that I have seen women gathering snails to make broth for the support of their families. It is true that the Cornish tinners have lately been very riotous and the gentlemen of the county have been obliged to call in the aid of the military, but when the extremely low price of Tin and the very high price of corn are considered, much may be said in defence of these poor industrious labourers.'

The riots referred to took place a little earlier when the price of grain had soared so that the poor could not afford it. Warehouses were broken into, the grain removed and sold at rock bottom prices. The military were called in and peace was restored. Such food riots had broken out sporadically throughout the eighteenth century, when the cost of essentials had shot up far faster than earnings.

The situation was getting desperate and Boulton and Watt needed to do something to protect their own concerns and they took 1/32 interest in several mines to inject capital to help keep them going. Mining companies began looking for ways to save money. One idea that appealed was to ask Boulton and Watt to give up their premium during the crisis. John Vivian of Poldice was one of those who put the proposal to the Soho company. Boulton's rejection of the idea was firm and reasoned, as he explained in a letter to Thomas Wilson, the Cornish agent, but he did offer one small concession at the end:

> 'JW wrote you last post informing you of the proposition made by Mr. Vivian & others that unless the Lords of Poldice & ourselves would agree to give up entirely all manners of dues on premium arising from that mine, until it should have repaid £7,000, which they say new engines will cost, they would vote for stopping the mine entirely. As they did not seem disposed to make the same demand in regard to the merchants profits we look upon the proposal as partial & unjust. And even setting that consideration aside we cannot agree to the proposal without receiving a far greater injury than the stopping of the mine can do us, by setting a precedent for demand which must end in the annihilation of our income, & therefore as we must make a stand somewhere it is best that it be understood by every body that we will upon no occasion agree to pay for or contribute towards Engines or those repairs or costs of mines. But in this present case as Poldice mine has not been profitable nor is soon likely to be so, on the current expences of working the mine independent of new erections, we will make them an abatement of part of our dues until the mine shall have repaid the losses she has made from the 1st of May last.'

The Birmingham company was inundated by other mines requesting the removal of the premium, and as Boulton wryly noted they were now regarded as 'oppressors & tyrants', and at one point they were even worried about Wilson's safety as their representative. He was advised to have his cases packed ready for a speedy getaway. It was never necessary. The crisis did pass. In 1782, the price paid for copper ore for the whole year fell to a low point of £152,434 but by 1794 – there are no statistics for the early 1790s – the price had risen to £320,875. It would continue to rise through the rest of the 1790s. The 1780s also saw a growing demand for copper from the shipbuilding industry. Wooden ships were plagued by marine life, from the ship worm, *Teredo navalis*, that ate into the hull, to barnacles and weeds that stuck to the wood and slowed the vessel down. It had been known since the beginning of the century that sheathing the hull below the waterline with copper solved the problem, but it was expensive. However, ships treated in this way could stay afloat longer without having to come in for repair and were faster and more manoeuvrable. It was the American Revolutionary War that decided the Admiralty in 1775 that the whole fleet should be treated.

The advantages were now so obvious that the merchant fleet followed the Admiralty lead. Vessels treated in this way were a better investment – they were literally copper bottomed.

Arguments over premiums were only a part of the problem faced by Boulton and Watt in the South West. Watt's patent effectively stopped anyone else from building an engine that used any form of condenser or air pump and the agreements between the company and the engine's purchasers also prevented anyone attempting to tamper with or make alterations to any part of the mechanism. This was not an arrangement that went down well with the mining community, who had been used to doing what they pleased. Several mine captains had made adjustments to their old Newcomen engines to improve performance, among whom was the captain at the great Dolcoath mine, Richard Trevithick. His main contribution was to completely redesign the boiler for an engine delivered in 1777. He was father to a son, also Richard, who was to become famous as the builder of the world's first railway locomotive, but was a constant irritation to the men of Soho, though he was by no means the only engineer who was considered to be pirating the Watt invention.

Trouble began with a claim from a dissenting clergyman Humphrey Gainsborough, brother of the famous artist. He was an amateur inventor, who had produced a number of ingenious devices and now declared that he had invented a steam engine on the same principles as Watt's. It too had a separate condenser and an air pump and his claim was strongly supported by the Cornish engineer Jonathan Hornblower. This was hardly surprising, as Hornblower was also working on a steam engine of his own and anything that brought the Watt patent into question could only make his life easier. Watt reacted angrily and Gainsborough, who everyone was agreed was a man of impeccable honesty, was understandably upset.

> 'Those who know both inventions have assured me that mine is totally different from your's. I must therefore leave you to act at your pleasure at the Patent Office, especially as it is impossible for me to be in town at present, and when God only knows, both I and Mrs. Gainsborough being very ill. As you have been ungenteel enough to give me unnecessary trouble, I am only sorry that I did not endeavour to hinder your Bill passing in any form, which I have good reason to believe would have been in my power.'

It must have been a worry for the Birmingham pair, since if Gainsborough could make good his claim, their patent could have been lost. In the event,

Gainsborough never pursued the matter. Hornblower, however, was to prove a much more dangerous adversary in later years.

Matthew Boulton was aware that he had to win the trust of the mine captains to receive orders for engines. He invited several of the leading men to Birmingham, including Richard Trevithick Snr. They were certainly interested in what Boulton had to show them – so interested, in fact, that when they left the factory, a set of working drawings went with them. They had managed to persuade someone at the factory to let them take the drawings back to their hotel, promising to return them the next day; they did not appear back at Soho. Boulton was not amused.

> 'We do not keep a school to teach Fire Engine making but profess the making of them our-selves. When you reflect on this matter you will readily perceive that there was an Impropriety in taking the drawings without our knowledge & consent and so I am persuaded you would not deviate from the Character of a Gentleman I doubt not that you will return the drawings with every copy that hath been taken and which I hear have been exhibited in Cornwall.'

Trevithick Snr soon sank even lower in Boulton's estimation when he had the temerity to criticise a new engine. He and his colleagues were now the 'Infidels of Dolcoath'. A second meeting in 1779 confirmed that relations could hardly be worse, as Boulton wrote: 'I was so confounded with the impudence, ignorance and overbearing manner of the man that I could make no adequate defence and indeed could scarce keep my temper.' Other Cornishmen were soon to incur his wrath as well.

Jonathan Hornblower had installed the first Boulton and Watt engine at Ting Tang mine. The company provided detailed drawings and instructions on exactly how everything had to be done. A set of instructions from 1779 run to many pages and 52 sections on everything from building the engine house to sorting out a botched job – 'If any of the joints are not at a proper angle fill them up with lead'. But Hornblower did what he would have done with any other engine – started doing things differently where he thought he saw an improvement. He was not only told to stop but was taken off the job that was then completed by Thomas Wilson. Worse was to follow. Hornblower began building an engine himself on a new principle; the first compound steam engine. In the existing engines, steam exhausted from the cylinder was instantly condensed. It was, however, still under pressure, so in the Hornblower engine the exhaust steam was fed instead to a second larger cylinder to provide extra power. His first Cornish engine was built in 1791 at the Tincroft mine. The high pressure cylinder was 21 inch diameter with a 6 foot stroke, the low pressure, 27 in diameter and 8 foot stroke. The two were arranged side by side, with the low pressure nearest the end of the beam. It was not a great success. Hornblower was still starting with steam at a comparatively low pressure, so that any advantage gained from compounding tended to be lost in increased friction. Hornblower went on to build nine more engines.

Compound engines were to prove of immense value over the years in many areas from working textile mills to powering oceangoing ships, but were comparatively unimportant in Cornwall. Hornblower took out a patent for the compound in 1781, but Boulton and Watt only became alarmed when they heard an engine had been erected at Radstock colliery in Somerset. Observers were sent to Somerset and Watt issued dire warnings of what would happen to anyone infringing his patent and, to make sure everyone knew what those items were, he published a list locally. But when he discovered that the engine was not performing well, he decided to take no action at that time, but to hold fire in case some more serious development followed. 'One should not warn a man that we mean to break his head, lest he put on a helmet.'

That was not the end of the matter. Hornblower asked parliament to extend his patent for the compound so that he could develop the idea, adding rather cheekily that he was encouraged to apply because Watt's patent ran for thirty years. He also claimed to have come up with the idea in 1776 – 'previous to his knowledge of any other kind than those invented by Mr. Newcomen'. Boulton and Watt opposed the extension, though they employed a curious argument. They had not attempted to try and stop the earlier experiments but did so now because he had built two engines. It hardly seems reasonable to let a man invest time and money in an invention and do nothing, then step in when he looks like proving successful. But the final statement in their evidence, bristling with italics and capitals, set out their precise objection: 'They, therefore, undertake to prove by competent Evidence, *that Mr. Hornblower's Engine is a direct and palpable PLAGIARISM of Mr. Watt's invention.*' In the event, the whole affair came to nothing, though it was significant that all the Cornish MPs backed Hornblower. But a more formidable opponent had appeared with a machine that definitely did work and worked well. This was Edward Bull.

The Bull engine was worked on a different principle. There was no nodding beam. Instead the steam cylinder was inverted over the pit, driving the pump rods directly. Wilson wrote to Soho to pass on the news about Bull's activity in 1791:

> 'We have heard that Bull has ordr'd his Engine that he has agreed to build in Givindrow to be cast at Harveys & that it is to be an invented one but on what principles is still a secret.'

Watt was not especially concerned as he had a low opinion of Bull's abilities and, had he known that what was planned was an inverted engine, would have dismissed the idea. He replied:

> 'In respect to Bull he must for the present take his swing, ten to one that he makes a very bad affair of it by dint of improvements & I mistake if his employers will have any money.'

It is not clear who Bull's backers were, but he certainly had no funds of his own. There were, however, adventurers prepared to finance engines that could be built without paying the Boulton and Watt premium.

Bull had another engineer to help him, the young Richard Trevithick – a new generation of the family to plague the Birmingham men. He was taking a risk, because there was every likelihood that if the Bull experiment failed, he would be out of favour with the most important engine builders in the world. Bull was already persona non grata so it made little difference to him. Trevithick, however, was to prove throughout his life that taking risks was second nature to him.

By the end of 1792, Bull was telling the world at large that his engines were superior to any others, and Boulton and Watt could no longer ignore the threat. Bull's claim that his engine was totally new and was not covered by the Watt patent rested mainly on the inversion and lack of overhead beam. This infuriated Murdoch, who put his hat on a table and asked, 'What's that?' and, receiving the obvious reply, turned it upside down and asked 'What is it now?' Bull suggested that matters could be settled by a trial between his engine and a Boulton and Watt of similar size, but the request was turned down – to accept the challenge would have given Bull legitimacy. Inevitably, there was rancour in Cornwall and once again the unhappy Wilson found himself at odds with his neighbours, who supported Bull. He received a sympathetic response from Soho:

> 'I am really very sorry for you that you should meet with such mortifications in our cause but we cannot remedy them & the best is to be patient, perhaps you have it in your power to piss upon *them* in your turn.'

Inevitably, the whole affair ended in the Courts in 1793 and the verdict went against Bull, but there was a sting in the tail in Lord Chief Justice Eyre's summary.

> 'But I think there is a question behind and that question is admitting it to be clear … that there has been a pirating of that invention – admitting it to be clear too that there has been Letters patent yet it is the Language of the Law that these Letters patent can be of no avail unless there is a true Specification of that invention. I confess I have myself very great doubt whether this Specification is sufficient.'

In fact, Watt's patent was deliberately set out in rather vague terms so as to cover as many areas as possible. The judge clearly had a suspicion that this was the case and that the wording might be so imprecise as to render the whole patent invalid. This offered an opening for anyone to bring a challenge. In that sense, it was a hollow victory for Boulton and Watt. There was another problem that now appeared. The injunction prevented Bull building his engines – but there was nothing said about Trevithick. Rather belatedly, Boulton and Watt realised that there was now Trevithick Jnr to contend with – a man about whom they knew so little that they had to write to Wilson to enquire what his first name was so that they could issue a writ. They got the writ, but serving it would prove far more difficult than they could ever have imagined.

Trevithick and Bull decided to install an engine at Ding Dong mine. This was a good choice for the mine was in a remote setting, and the engine house can still be seen on the open moorland overlooking St. Michael's Mount. The approach was by rough tracks and anyone using them would be seen long before they reached the site. The bailiffs came when no one was present, but writs have to be served in person. John Bolitho, whose father had worked at Ding Dong, tells the story of what happened next.

> 'Boulton and Watt came down with an injunction printed out and posted it up on the door of the engine-house, and upon the heaps of mine-stuff, and nobody dared to touch them. But Captain Trevithick did not care; he and Bull and William West came and turned the cylinder upside down, right over the pump rods in the shaft; they took off the cylinder top (it was the cylinder bottom before they turned it upside down).'

Trevithick was obviously furious at having been singled out and took out his bad temper on Wilson and in Wilson's own words 'bespattered me at a terrible rate'. The Soho lawyers could have avoided the necessity to pursue Trevithick if they had been able to prove that they were in an official partnership, but neither man seems to have been a great one for official agreements and formal documents, so there was no more than hearsay evidence for the connection. It seemed that they would have to corner the man himself, somehow or other. Finding anyone prepared to take on the task proved tricky, given Trevithick's reputation as a man of great strength and short temper. As a young man, Trevithick had shown off his strength by throwing a sledge hammer over the roof of an engine house. One bailiff did approach the mine, but he was grabbed, tied up with rope and dangled over the open shaft and politely asked if he was sure he wanted to serve the writ. He felt that perhaps, on the whole, he'd rather take it away again and the sooner the better.

There were more angry exchanges of words, but no action; writs remained unserved and work on engines continued. Then Bull and Trevithick decided to walk into the lion's den. They were on their way to Coalbrookdale to discuss engine castings and stopped off in Birmingham. They were recognised and found in a pub on the evening that there was a fireworks display at the Soho works to celebrate a victory over Hornblower. Now there was a double cause to celebrate; Trevithick finally received a writ. The battle was effectively over. There was nothing to be done – a premium had to be paid to Boulton and Watt for the engines they had installed. However, by 1796 it became clear that the Bull engine was not quite as good as its inventor claimed and that Boulton and Watt had nothing to fear from competition in that quarter.

The next few years saw Trevithick turning to other ideas for devices to serve the industry, with rather more success than he had with the Bull engine. He had a fresh incentive to do well, for he had married Jane Harvey of the highly successful Harvey family who had ironworks at Hayle in 1797. That year, he developed the plunger pole pump. It was not a new idea, but he greatly improved on earlier versions. Most pumps at the mines were bucket pumps, in which the work was done by sucking up the water on the upstroke. In the new pump, the plunger forced water up through an H-piece into the pump column. The water was carried to the surface in a series of lifting operations, typically 30 fathoms at a time. The device saved fuel, because the plunger fell under gravity and power was only needed to raise the plunger, not to carry up water as well. It was widely adopted.

Trevithick had not, however, abandoned the idea of developing steam engines. He began to question whether an engine needed to have a separate condenser at all. What would happen if he dispensed with it altogether and compensated for the loss of vacuum pressure by using high pressure steam? He was no scientist but he had a friend who was, Davies Gilbert. He sent him the question what would be the loss of power if he applied steam at several atmospheres and instead of condensing it simply let it exhaust into the atmosphere? Davies replied that the loss could never be more than one atmosphere and added that such an engine would not require an air pump. As Gilbert recalled, Trevithick was overjoyed; with no condenser and no air pump, he could not be charged with piracy and could build engines with impunity.

The new engines soon went into production, and became known as 'puffers', because a puff of steam escaped from the machine at each stroke. The first engine was set to work at a spot Trevithick knew well – Ding Dong – and was so successful that over the next four years, at least fourteen engines were ordered. These were all whim engines, used for hauling material up and down the shaft. Boulton and Watt, unable to sue, set about denigrating the new machines and set up a trial which they supervised, pitching one of their engines against a puffer. Not surprisingly, since they had both set up and run the trial, early results looked bad for the puffers. Then Trevithick came along to take charge of his own engine and discovered that somehow or other the piston had now become half an inch shorter than it should have been, resulting in a loss of power. Trevithick was now confident of success and took a wager of £50 that his engine would win the contest. Benjamin Glanville was in charge of the Boulton and Watt engine and held the wager. A new piston was installed in the puffer and the contest got under way and according to an eye witness account, the puffer beat the Boulton and Watt 'all to nothing'.

Trevithick was aware that the old boilers then in use, which were often little better than oversized kettles, were no longer adequate. The Boulton and Watt engines were typically supplied with steam at no more than 3psi., where Trevithick was looking at pressures up to 50psi. He developed a round boiler with an internal U-shaped flue. A firebox was set in one end of the flue and the hot gases passed down the flue, heating the surrounding water to arrive at a chimney next to the firebox. Exhaust steam from the engine was passed up this chimney, which acted like a blast pipe, sucking air through the firebox to increase the heat. Inevitably, the pipe became very hot, but that heat was not wasted. The water used to cool the pipe was fed back, already warm, into the boiler. The Trevithick boiler was so widely used that in time it came to be known simply as the Cornish boiler.

Trevithick soon realised that with high pressure steam you no longer needed to build bigger engines to get more power – you could increase the pressure instead. He then realised that if engines could be made small enough, they could be put on wheels and towed to wherever they were needed. From there it was a leap of imagination – make the engine turn the wheels, so that the puffer could move itself. The stationary engine would become a steam locomotive. But that is not part of this story. His first road locomotive trundled up Camborne Hill at Christmas 1801. By that time, the Boulton and Watt patent had been overthrown. The courts agreed with the judge in the Hornblower case that the patent was too vague and ended it. The way was open for a new world of engineering in the mines of the South West.

Chapter 7

THE HEALTH OF THE MINERS

Although by the end of the eighteenth century, whim engines were available, they were not used to help the men go up and down the shafts. In many cases, ladders were placed in working shafts, so that the workers had to climb up and down to the accompaniment of the steady rise and fall of the pumps. Astonishingly, there were some intrepid tourists who were so keen to see the workings of a mine that they were prepared to undergo the hazardous journey. One of these was James Forbes who, in *A Tour into Cornwall*, 1794, gave a graphic account, not only of the climb but of the many other difficulties they encountered in moving around the mine. Few modern tourists would welcome the opportunity to follow in his footsteps. Having got himself kitted out with a woollen shirt, trousers and what he described as a 'nightcap' and been given his supply of candles, he was ready to be led underground by the mine captain. He was then taken to the top of the main shaft:

> 'It requires a strong stomach and a large degree of curiosity to go through all this – for besides the fatigue and toil in the mine, the cloaths they give you are greasy and filthy to a degree, smell abominably and are often stocked with a republic of creepers. A miner went first to serve as a guide and to caution us against the danger which frequently arises from the broken staves in the different ladders, for the Captain told us that if we made a false step to one side or the other we should be ground to atoms in the steam-engine or dashed to pieces in the shaft. The descent resembled a large well with an immense machine for the purpose of draining the water, in motion all the way down. We continued to descend ladders which were from four to five fathoms in length, and being soon wet through, weak from want of proper respiration and half-stifled with the fumes of sulphur, began to hesitate whether we should proceed or not. I had no idea of the difficulty and danger attending such an undertaking and only wonder that accidents are not more frequent among the miners, who run up and down these slippery places like lamp-lighters, singing and whistling all the way. At about eighty fathoms depth we came to a vein of copper-ore where some poor creatures were

> busied in the process of their miserable employment – with hardly room to move their bodies, in sulphurous air, wet to the skin and buried in the solid rock these, our fellow-mortals, live and work for their daily bread, pecking out the hard ore by the glimmering of a small candle whose scattered rays will hardly penetrate the thick darkness of the place. Proceeding in our descent, we reached at length the bottom of the mine and stood 139 fathoms below the surface of the earth. In this mine is a vein of tin also, and a communication is dug from the Copper to the Tin. Through this we crawled upon our hands and knees and afterwards had to cross through a rapid stream whose waters rushed abundantly over us, as we crawled along in space just sufficient to admit us upon all fours. Working our way thus in a direction of north to south, we came at last to the shaft of the tin-mine. Here we saw two figures that hardly wore the appearance of human beings, *singing at their work.* We find it exceedingly difficult to pay them a visit, as we had to descend by a single rope down a chasm, never broader than a chimney, until we reached the Load where the miners were employed.'

He then had the long climb back to the surface, and by this time felt quite faint and declared that he could hardly describe the luxury of being able to breathe fresh air again. He reported that at the bottom of the mine, the heat was so intense that men worked naked and that it made no difference to them whether outside was ice and snow or a blazing summer sun.

In many ways, this account is similar to that given by the young boy describing his first day down the mine, but it is worth quoting because it shows just how exhausting it was for a healthy adult to make the journey down what was by no means the deepest of mines – and he didn't have to take equipment down with him. It is also interesting to discover that even after such an arduous journey, moving from one part of the mine to the next was equally unpleasant and tiring. Yet he describes the workers as being cheerful and agile. But over the years, more and more evidence was gathered suggesting that the men who worked down the mine would suffer from their efforts in later life.

L.L. Price, in his book *West Barbary,* 1891, noted that the mortality rate among the tin and copper miners was worse than that in any of the coal mining districts. Yet in the collieries, there was a horrendous loss from explosions, accidents in which hundreds could die at a time, while such accidents were virtually unknown in the South West. The Royal Commission on the Housing of the Working Classes 1885 compared mortality rates over long periods of the nineteenth century. What emerged quite clearly was that though the death rates in colliery and tin and copper mining areas were roughly similar for young men, when they approached middle age the Cornish miners' chance of survival quickly diminished. So that in the age group 45 to 55, for example, the deaths per thousand of population were averaging 33.51 against 16.81 in the colliery districts and 14.76 for males in Cornwall as a whole. Price then set out what he believed were the causes.

Various reports set out the causes of death, which were chiefly down to lung and respiratory diseases. They were also reported as suffering from a variety of digestive complaints, but the Commission was keen to point out that this was not due to alcoholism:

> 'The miners as a class are well conducted and temperate. Large numbers have taken the pledge, and kept it, and, whatever may be the causes of the disease to which they are liable, the habit of intoxication cannot be assigned as one of them.'

This sobriety can be traced back, in part, to the spread of Methodism in Cornwall, which was said to have a higher proportion of the population following the Wesley doctrines than any other county in Britain. It was not always so. Richard Polwhele in his *History of Cornwall*, 1817 edition, wrote: 'The ardent spirits to which the vulgar Cornish (and the not so vulgar, it might be added) are habitual, and unquestionably injurious to health.' There were generally 'winks' or pubs that had never troubled the licensing authorities for any sort of permission to serve alcohol, near many of the mines. They became less common in later years. So, if drunkenness was not an issue and accident rates were comparatively low, what was the cause of mortality among the miners? A Report of 1842 into the miners' health charted the development of what was known as 'miners' consumption' in sombre detail:

> 'A degree of shortness of breath is the first symptom complained of, and it is long whilst itself gradually increasing, unaccompanied by any other. The digestive powers are unimpaired: the miner says 'his heart is good' and when seated feels as if his health were perfect. After a time, however, he begins to lose flesh, and his general strength slowly fails … At length cough and expectoration moderate in extent come on, and hectic fever is at last established with the usual concomitants though the progress of each stage of decline is still slow, and marked by no urgency of symptoms.'

The symptoms may have lasted anything from two to ten years before the unfortunate man finally succumbed. It was clear that, with the condition, limited almost entirely to the mining community, it must be something in the nature of the work or the working conditions that made it so prevalent.

Here we come back to the account quoted at the beginning of the chapter. While the gastric disorders could be put down to a poor diet, the respiratory problems were caused by climbing the ladders and in particular by the young men who made things worse, because the effects were 'exaggerated by the indiscretion of the miners themselves, specially the younger me, in mounting the ladders in too much haste'. It was, in this view, favoured by the authorities, all the fault of the men themselves. Price would have none of this. The problem was the system itself and the working conditions in the mines. It was clear from the earlier description that it was not merely the exertion of using the ladders, but the fact that in descending, the men were frequently soaked to the skin before they even started work. The Royal Commission into mines chaired by Lord Kinnaird in 1864 recorded cases of mines where the ladders went down to a depth of 280 fathoms, where just going to and from work took three hours. To put that in context, prisoners sentenced to hard labour and sent to work on a treadmill were not allowed to do more than the equivalent of 1,200 feet per day. And once the men reached their workplaces, they were often damp and poorly ventilated. The conclusion was that 'the exposure to cold and damp at the surface or in the shafts, and other parts of the mines' was 'the most frequent exciting cause of disease operating with peculiar severity among men in a state of perspiration or exhausted by working in close places, or by ascending ladders from great depths'. The Commission was critical of the owners for permitting the system of ladder climbing to persist. If the mines could be supplied with steam power for lifting out the ore, why could it not be supplied everywhere for moving the men? Lack of technological ability was certainly not the answer, for by this time Cornish engineering had advanced at a rapid rate. There was an answer, which we shall come to shortly after taking a look at how engines developed in the nineteenth century.

With the Watt patent no longer valid, the way was open for Cornish engineers to develop engines in any way they thought fit. A great impetus to improvement was the publication of regular reports on engines and their performance. It began with the appearance of the *Engine Reporter* in 1810. A year later, Joel Lean was appointed as Registrar and Lean's *Engine Reporter* continued to appear throughout the rest of the nineteenth century. It listed the engine, location, cylinder diameter, stroke in pump and duty. The first issue named just eight engines: three at Wheal Alfred; three at Dolcoath; one each at Cook's Kitchen and Wheal Fanny, with an average duty of 15.7 million. Now that figures were available and everyone could see just how efficient different engines were, there was an incentive for engineers to see their engines at the top of the list. Cornwall became a battleground for a duty war.

A great deal of effort was put into improving the power of the great pumping engines. Arthur Woolf was one of the pioneers of compound engines and, in May 1811, he put an advert in the *West Britain* claiming that his engine 'would require not more than One-third part of the Fuel employed in working engines on Messrs Boulton & Watt construction'. He was then engaged by the United Crowan group of mines and began erecting his first compound. The other approach was to simply increase working pressure. Richard Trevithick led the way with a 24-inch engine at Wheal Prosper. This was an inverted engine, which is more or less what the name suggests. The cylinder was turned upside down, with the beam below the piston instead of above it. This was worked at a pressure of 40psi, the highest yet for a beam pumping engine. At this stage of development, however, there was still a difficulty with raising pressure safely in a boiler. When Trevithick's friend Davies Gilbert stood on one of the engineer's new boilers and was told the pressure, he declared that it was probably the nearest he had ever been to death. Over the next few years, bigger and more powerful engines were built, working at ever increasing pressures. In 1823, the first of many 80-inch engines was erected, a size that was soon to become popular for big pumping engines. In 1824, William Sims installed an 80-inch at Polgooth, which was soon recorded with a duty of 43 million. But size was not everything. In 1825, a more modest 60-inch engine at Wheal Hope built by Samuel Grose was recorded at 45.2 million. The secret was that Grose took great care in the construction details to ensure that when steam was raised it was used as efficiently as possible. There was no shortage of expertise in Cornwall and the technology of the steam engine was being constantly improved, but few it seemed were interested in applying their ingenuity to the deadly work endured by the miners in their long, daily climbs.

In 1833, a new device was installed by Herr Doreell at a mine in the Harz Mountains in Germany. It was comparatively crude, but the principle was sound. Just as pump rods go up and down with each swing of the overhead beam, so would a similar structure in this engine, the difference being that it would have some form of protruding part attached, that a man could stand on. Then small stages had to be built into the side of the shaft. A man ascending would wait until he could stand on the moving part, and at the end of the upstroke step off onto the platform. Then he would wait for the next stroke and repeat the operation until he finally arrived at the surface. The German engine was said to have used old wooden pump rods, with huge nails hammered in for the men to stand on. It was called a *fahrkünst* or travelling shaft, but in Britain became a man engine. The original German version was powered by a water wheel, but a steam engine could easily have done the job.

News of the invention reached Britain and the Cornwall Polytechnic Society set about promoting its use in 1833. The Society, which was soon to become the Royal Society, was set up by the Fox family, the most famous of all Quaker families. Not surprisingly, as well as encouraging advances in technology, they were keen to promote the welfare of the workers. Initially, their efforts had little effect. The Society circulated details of the German machine among the leading adventurers but no one responded positively. Then, by 1838, when nothing happened, they offered a subsidy to anyone prepared to install an experimental engine, based on the German model. Three mines replied, but two rapidly withdrew because there did not seem to be an appropriate shaft available for the enterprise, while the third said it would take eighteen months before they could begin work. Finally, the Tresavean adventurers decided to install a small water-powered man engine at their mine in 1842. It only went to a depth of 24 fathoms, but the experiment was a success. Two months later, the engine was extended to a depth of 290 fathoms below the surface and the water wheel replaced by a 36 inch steam engine. The man engine had two eight inch square rods, and they raised the men in 12-foot stages. It was an undoubted success, but there was no rush to build more. It had been quite expensive to install, and many miners distrusted it, some referring to it as 'a machine for killing men'. But the main problem was the reluctance of mine adventurers to make the investment. As late as 1880, the Inspector of Mines in his official report wrote of the 'disgraceful supineness' and 'surprising apathy' of the adventurers. In general, the man engine was a great blessing to the mining community – at least for those fortunate enough to work in a mine that provided one. It was in fact shown that installing a man engine could increase productivity, simply because it led to a more efficient, healthier workforce. It was such a boon to the miners that it was even celebrated in verse by W. Francis, a poet from Gwennap:

'The engine by which he is eased from below
Now supersedes climbing, health's deadliest foe –
This miners know well and their gratitude show.

Their core being o'er from labour they cease,
And delighted avail them, O Loam of the ease
Thy genius procured them and joyful ride
On the rod, while others descend by their side.'

Loam mentioned in the poem was Michael Loam, who had installed the first engine, and is widely credited with inventing the device – the German predecessor having been conveniently forgotten. It was never quite as useful as the poet describes. In many mines, where man engines were installed, they failed to reach the bottom of the pit. At Dolcoath, the engine went down to 240 fathoms but the bottom was at 352 fathoms, still leaving 112 fathoms of ladders. Other pits were worse. At Carn Brea the engine only reached 170 fathoms, half way down the shaft, and at Cook's Kitchen the engine only went down a third of the way. That was not the only problem. The man engines could, sadly, become 'killers of men'. In May 1863, a miner tried to carry two water kegs down with him, which was against regulations. Trying to balance then proved too difficult, he lost his footing and fell to his death. An even more tragic case occurred that same year when a boy, who had used the ladders to climb down, decided to try the new machine for his ascent. He only got as far as the first staging, when it seems he was paralysed with fear and instead of moving upwards remained still, only to be crushed when the rods moved again. These were avoidable accidents, but there was an inherent danger, caused by the slippery mud and clay that stuck to the steps. There was an average of two deaths a year on man engines in the mining districts. But it was not until the early twentieth century that there was a major accident.

The engine at the Levant mine had first been installed in 1855 and gradually improved in efficiency over the years. By the time of the accident in 1919, it was working down to a depth of 266 fathoms. The timing could not have been worse, happening just as shifts were changing, with one set of men starting up and the other shift starting to descend. One of the connecting rods broke, sending the main engine crashing down. Most of the damage happened at the top of the shaft and it was here that 31 men were killed and 11 seriously injured. The remainder of the 150 men in the shaft at the time escaped with only minor injuries. However, the men below ground were now trapped, and the miners on the surface at once got a rescue under way, and other miners from nearby mines joined in the effort. It was dangerous work, and it took many days before everybody, including the dead and the injured, could be brought to the surface. It was later discovered that the accident had happened due to a faulty iron plate that connected the steam engine to the rods. It was the worst single accident in the history of tin and copper mining. A part of the man engine can still be seen beside the Levant whim engine house on the cliffs near St. Just.

Climbing was not the only cause of ill health described by Lord Kinnaird's Commission; they also identified bad ventilation as a cause of respiratory diseases. There is no shortage of accounts of just how bad conditions could be. One miner in the St. Just area described how they reached a level where the air was so bad that the only way in which they could keep their candles alight was to have a small boy constantly wafting fresher air over them. Worn out by this constant fanning, the poor lad fell asleep, and the first the men knew about it was when all the candles went out. They found it impossible to relight them and had to stumble their way out of the level in pitch darkness. Air conditions were especially bad after blasting as was described by a seventeen year old boy working at the Fowey copper mine in 1842:

> 'The air was poor where he then was and he had a pain in his head after working some time, which lasted for hours after he came to the surface. Almost every morning he had a cough and brought up some stuff black as ink. In the place where he was working they used to 'shoot' [blast] three or four times a day, after which they could not go into the end for half an hour, as it was full of smoke. He would eat his pasty in the level, where there was better air. Though he sweated a great deal and was very thirsty he could not generally get water underground.'

The 'black stuff' obviously came from the particles in the air following the explosion. No doubt he was worried about what he coughed up, but, as an old miner explained to me some years ago: 'It's what you don't cough up that will kill you'. Conditions in the working environment were often atrocious and life at home was not always that much better.

Chapter 8

HOME LIFE

Unlike coal mining areas, there was very little in the way of company housing in the South West. Tin and copper mining was speculative and a remote mine might well turn out to have little value so there was no point in spending money on houses for a workforce that might not be needed after a time. Often it was left to the miners to build houses for themselves. Because of the nature of the landscape in Cornwall and Devon, with their large tracts of moorland and stony wasteland, there was land available that no one really wanted or could use. The stones that made it unsuitable for agriculture were there for the taking as a building material. There are very few records to indicate the type of houses being built by miners in the eighteenth century, but R.Q. Couch in his *Statistical investigation of the Mortality of Mines*, 1856, gave this account of housing as he found it, and it would certainly have been no better a century earlier.

> 'The population is mixed but the miners invariably occupy the most exposed and worst built cottages … surrounded by cess pools, broken roads and pools of undrained rain. The village of Amal-Voen is like a cluster of cottages huddled together on the top of a hill with scarcely space between them for access. The bedrooms are rarely more than one in each house, and open to the ceiling. This gives that appearance of space; but if the roof is slate it produces great heat during the summer days.'

When researching for my book *Remains of a Revolution* in the 1970s, I investigated a pair of cottages, then in a semi-derelict condition, on the approach track to Botallack mine. Both were constructed out of a jumbled mass of granite boulders of all shapes and sizes, with single doorways, and miniscule windows with frames constructed out of what appeared to be timber cast-offs. The larger of the two had a central doorway, leading into a single room with a rough, stone clad floor. There was an equally rough fireplace at one end, with a large, straight chimney open to the sky. Upstairs was another single room, below the rafters and covered by a stone slate roof.

The second cottage was single fronted and slightly smaller. A lean-to at the back of the larger cottage showed every sign of being a later addition. These primitive cottages were apparently grander than many built in the eighteenth century, which would have been similar one up one down but with only an ash floor and crude thatch instead of slate on the roof. There were a few speculators who built cottages and there are examples dating from around 1800 near Hayle. A curiosity of this area was that, being near smelting works, a cheap building material was available in the form of blocks of slag. Cottages of this type can still be seen at Gwithian.

The work of clearing the ground was often arduous. Richard Thomas in his *Report on Mining Area,* 1819, described the process for Carn Marth, a hill near Redruth:

> 'Thousands of acres of down, commons, and wastes, have been enclosed and are continuing being enclosed by the miners and others on a small scale, generally from three to six acres in a tenement, on each of which one or two cottages are erected. The soils on the granite hills are generally better than on the killas [a form of sedimentary rock] downs, especially the grass but in many situations cost immense time and labour in clearing away the rocks and stones. This has been particularly so the south-west slope of Carn Marth, where the enclosing tenants (chiefly miners and labourers) have had to blow the rocks to pieces with gunpowder before they could remove them to where they are collected and heaped up to form fences to their little plots. Two or three thousand tons per acre are then removed from some spots before the ground is cleared. About thirty tenements of this sort are enclosed on the south and west sides of Carn Marth, containing about 150 acres all together. Many more would have been enclosed had not the land been divided among so many owners. The cost of so many leases, which the tenants have to pay, amount to more than the freehold of the surface is worth without reckoning the circumstances of being compelled in some cases to attend the Lord's Court at loss of time and labour or else to pay a fine.'

The usual arrangement with the landowners was that the miners, as well as paying rent, had a lease that expired after three generations. In fact, the landowners seem generally to have charged what the Report referred to as a 'high rent' of several shillings per acre, and when the 'three lives' were up, the family had to pay anything from £25 to £50 if they wanted to continue living in the property. The miners had all the trouble and expense of clearing the land and building their homes and at the end of the day still did not own them outright. This was obviously a good arrangement for the landowners, who had previously had land that was considered worthless and now had an income without having to exert themselves beyond signing documents and collecting ground rent.

Not all the miners lived in isolated communities, with cottages built on wasteland. There were also developing mining towns, such as Camborne and Redruth. The 1864 Commission provided a good deal of information about the development of Camborne in particular. At that date, one witness estimated that there were some 800 houses in the town lived in by miners and which they had built themselves, usually on a lease from the landowner that, like those on the moorland, lasted for three generations. The building costs were higher than in the country, generally between fifty and eighty pounds.

Many of the houses were built by tributers, who rarely if ever took out loans to cover the building costs. This was one of the advantages of the system. Although there were periods when incomes were low, there were also bonanza times when they suddenly found themselves making good money, perhaps as much as fifty pounds in a month. According to the 1864 Commissioners, this happened to most men at least once or twice in their working lives. Then, if they were prudent, they had the money to build. It was, of course, quite a big 'if'. The Camborne houses were generally regarded as having been well built, some had as many as six rooms, but these larger houses were invariably home to two families. Most had small gardens for growing vegetables.

Houses might be built, but all too often they were lacking even the most basic amenities. In the country, many cottages did not even have a privy. Drains were unknown and water was so scarce that in some cases it had to be fetched from distant streams. And in one case, this involved going to streams almost three miles from the village. Many women used to take their laundry to the mines and used the hot water from the steam engines to wash their clothes.

Not all mines were situated in areas where stone was freely available. An alternative was cob. This is a mixture of subsoil and straw, often with the addition of sand to ensure the right consistency. It is mixed with water, usually by trampling underfoot. A course of the wet material is then laid on a stone foundation. Once a course has dried out, the next can be added and so on until the walls reach the required height. Most cob houses had thatched roofs. The walls were generally about two feet thick, and provided excellent insulation, keeping the houses both dry and warm. As one miner reported, you could get up in the middle of a rainy night and the inside walls would be so dry you could strike a match on them, The other great advantage was that they were cheap to build. Hamilton Jenkins quotes one old miner who, on being quizzed while building his cottage, agreed that it was an expensive business, and when asked how much he reckoned it would cost, replied: 'Why, I reckon, sir, by the time I have turned kay in the door, 'twill have cost me fifty shellan.'

Going home after a long, hard shift should have been a great relief, but getting between mine and house presented its very own problems. First, the men had to change from their work clothes into their everyday wear in a room with few facilities and then make their way to a house that might be some distance away. In winter, especially, this might often mean walking in the dark, with nothing to light their way but an improvised lantern with a single candle in it. Apparently, a popular lantern was made out of an old, worn out boot. The conditions they met when they reached home varied from family to family.

The Cornish miners generally married young and they were known for a common practice in the South West, known as 'bundling', which was a polite way of saying that they normally had sex before marriage as a trial run. They also had children at an early age and many of the young women, faced with the living conditions and hardships they had to endure in so many cottages, must have struggled to cope. But the 1864 Commissioners had little sympathy with those whose standards fell too low. They quoted a witness:

> 'You will find two men and one has got a clean, decent, wholesome, industrious wife and that man's children will be kept as clean and comfortable as possible. You will then see one of the same 'pare' who has got a dirty, careless wife, and that family will be in rags, and yet that man will make the same earnings. One man will be well off and the other always in misery.'

A number of authors quoted cases where the women had allowed the home to deteriorate to the point where it was scarcely habitable. The assumption was always that the wife was to blame. No one seems to have enquired into the circumstances, and even if the husband had been an honest, sober, gentle man, the actual living conditions were atrocious by modern standards. It is hard to imagine the effort involved in trying to bring up a large family in the cramped surroundings of a two-room cottage with a total lack of facilities. And although the Commissioners dwelt at some length on the poor conditions of some houses, the Commission of 1842 reported that it was very rare to find such squalid conditions in the homes of Cornish miners. It would seem that the 'sluttish housewives' were, in fact, rare and the vast majority of the women managed wonderfully well in the most difficult circumstances.

The wife would have a hot meal ready for the miner when he returned after work, generally consisting of salted fish and potatoes, washed down with tea. When money was scarce, tea was simply too expensive, so the wife would go out in the hedgerows to pick the common herb, mugwort that would then be dried and used as a substitute. They would probably be astonished to know that in the twenty-first century, mugwort tea had become fashionable and was now more expensive than regular tea. If things were going well, the fish might be replaced by bacon or some other form of meat, but it was very far from being a feast, especially if the miner had a large family to feed. It is quite understandable that children were encouraged to go out to work as soon as possible, to help pay the growing food bill. Often, the men would stint themselves to ensure everyone got enough. One man who had endured five years of illness was helped out by being sent to the inn where he was given beef and beer to restore his strength, because if the food had been sent to his home, he would have felt compelled to share it out.

Things reached desperation point in the 1840s, which would become known as 'the hungry forties'. The problem began with a slump in trade in 1839, with widespread unemployment and matters were made worse as that year, and the next two years, had extremely poor harvests. That was not the end of the problem; two absolutely basic commodities became both scarce and expensive. The price of corn was kept high thanks to the Corn Laws, originally passed to prevent the import of cheap corn from Europe but retained to ensure good profits for landowners. Then potato blight devastated that staple crop. The effect was felt everywhere in Britain, most famously in Ireland, but the South West of England was certainly not exempt. Thomas Oliver was a miner who wrote his memoirs, eventually published in 1914 as *Autobiography of a Cornish Miner.* In it he described his own experience of those years.

'My father had the standard wages for surface hands, which was £2 5s a month and I was earning 10s. a month, so that £2 15s. a month had to provide for five of us. For our breakfast we had barley gruel, which consisted of about three quarts of water and a halfpenny-worth of skimmed milk, thickened with barley flour, a concoction which went by the name of

> 'sky-blue and sinker.' We lived about half a mile from the mine and I had to go home to dinner. I can assure the reader that I was sometimes so feeble that I could scarcely crawl along. For dinner we had sometimes a barley pasty with a bit or two of fat pork on the potatoes, and for supper a barley cake or stewed potatoes or turnips with a barley cover. Everything was very dear; groceries such as raisins and currants were 10d. per pound, tea was 4s. a pound and the common brown sugar 5d. a pound. I never saw at that time such a thing as jam. Barley flour was £2 per bag of 240 lb., and wheat flour £4 per bag of 280 lb.'

Those were the worst of times, but it is clear that for most of the nineteenth century, the home life of the miner was frugal, living in cramped conditions and surviving on a generally poor diet, only alleviated when the miner was fortunate enough to hit one of his rare bonanza times. Not everyone in the industry suffered in the same way. Mine captains were very much better off and there was no question of them building their own houses. The Trevithick family home at Penponds, for example, has been preserved. It is a substantial stone-built two-storey house with a thatched roof. Inside are two main ground floor rooms, separated by an entrance hall, with a large kitchen to one side and a handsome wood-panelled living room to the other. Other mine captains enjoyed even grander houses. In 1866, a house was built for the Duke of Bedford's mine captain James Richard in a magnificent situation overlooking the Tamar valley near Tavistock. A measure of just how splendid this building is can be gathered from the fact that today it is the luxury country house hotel, The Horn of Plenty. It is a far cry from a two -room cottage on a desolate moor.

Children's lives, for the few short years before they went to work, were spent, in part, in preparing them for what would be their new occupation. The 1842 Commission explained how this rather grim process worked for young boys:

> 'The parent … will convey to them a reflection of the sharp life which his hazardous and varied life has impressed on his own mind; and next by their own experience. Many years prior to his being sent to the mine the little boy is employed in ways which mimic his future occupation; with a shovel of a size suited to his strength he is employed in collecting from the road anything which may serve for dressing; this is placed in a small barrow, which introduced him to the work of wheeling; a step further in the same practice is afterwards made when he is busied in bringing water in a small

> cask from the well to his home. Other small tools, similar to those used in the mines, are often given and he may be seen at work with them in imitation of what he has noticed when he has been taken there.'

Some, but certainly not all, children went to a day school, for which the parents had to pay anything from a penny a week to 5s 6d a quarter. The only free education was supplied by the many Sunday schools run by the Church and the many Methodist and Baptist chapels. Although many children went to these schools, others from the poorest families often were kept away simply because the parents felt they could not afford to dress them in suitable clothing. The Church of England schools offered a very limited education, designed mainly to indoctrinate them into church practices. So they had Bible readings and were taught the Creed, the Catechism and the Commandments, but even in this limited role they proved unsatisfactory and the Report is highly critical of the institutions:

> 'The total inadequacy of the instruction which the Sunday school is capable of affording to the fulfilment of the ends of education, is admitted, without any exception, by the parties, almost all of them clergymen or dissenting ministers.'

The Church of England was perhaps the least successful in furthering education as it seems to have had little appeal for many of the mining families. In the eighteenth century, when livings were bought for younger sons of the wealthy and incumbents might have little relish for the work, it was not uncommon for parsons to pay scant attention to their congregations, though obviously there were notable exceptions. Churches were established in the larger towns and the more scattered mining communities might feel little connection with the religious life. All that changed dramatically, when John Wesley made the first of his thirty-three visits to Cornwall in 1743. He came from London with two companions and just one horse between them, and after crossing Bodmin Moor, Wesley's two companions stopped at the house of a stone mason, Trewit cottage, where they were offered food and drink and hay for the horse. Apparently, the owner of the cottage, Digory Ishell, was astonished when they dropped on their knees and prayed without using a prayer book. When, three weeks later, Wesley himself joined them, three hundred turned out to preach.

The appeal of Methodism for the community is not difficult to see. It had none of the ritual elements of the established church; the preachers spoke the language of the ordinary working people and they preached a simple message – if you have faith, you will be saved for all eternity. In the early years, all the preaching was done out of doors, and the most popular meeting place was Gwennap Pit, a simple amphitheatre with tiers of grassy seats. It was estimated that when Wesley preached here, 32,000 turned up to hear him. Methodism became widespread throughout Cornwall and at its peak in the 1870s, chapel membership was almost 40,000. The effects on the local population were considerable, not least since one of the key features of the religion was the denunciation of all alcoholic drinks. Cornish miners did not become sober teetotallers overnight, but where the religion was strongest, drunkenness became less of a problem, and the health of the workers benefitted. Chapels were built throughout Cornwall, but although their effect on the health of the community might have been largely beneficial, their contribution to childrens' education was minimal.

The day schools were no better than the Sunday schools, and again the Report is clear about just why they were so inadequate. This is a statement from St. Ives:

> 'Of the several teachers connected with schools in this district not more than three are considered fit for their functions. Some of the remainder are grossly ignorant of elementary education, and have probably reverted to tuition as being easier than manual labour.'

It was all too typical of the many statements given to the 1842 Commission.

Yet, in spite of these inadequacies, many commentators noted that the miners were better educated than most other working class people of that time. Partly, at least, this was because of the enthusiasm of the children themselves. Dick Hampton, known as the Cornish Pilgrim, became a well-known preacher, and in his autobiography, *Foolish Dick,* he described his own schooling. He was sent to school at the age of eight in the 1790s for which his parents paid 3d a week. Even this modest amount proved too much for them and they had to take him out of school after just seven months. But in that short time, he was taught to read a psalter. From this modest start, he persevered with teaching himself, until he was able to read the Bible.

Comparatively little is said about the lives of girls, but one doctor reported somewhat disapprovingly that girls under eighteen showed a 'passion for dress'. He was particularly critical of girls going out in thin shoes and stockings in unsuitable weather. And he noticed that girls who appeared at his surgery with complaints that could be put down to poor diet, still managed to find money for pretty clothes. Which goes to show that teenage girls in nineteenth century Cornwall were not so very different from the teenagers of today. The doctor's disapproval was generally shared by many of the upper classes in Cornwall, who laid the blame on the packmen, travelling salesmen, who tempted the girls and women with fine clothes, and allowed them to pay in instalments, but who were remorseless if the payments were not kept up.

Visits to the doctor were generally only paid for when they involved injuries at work, where the men had paid into a fund. The rules for the Wheal Buddock Club, dated 1839, laid out charges at twelve pence a month for men and sixpence for boys. As well as paying for the doctor or surgeon, there was also a living allowance of twelve pence a day for men and sixpence for boys, but there was also an ominous proviso – provided there was enough money in the coffers. There often wasn't and that was not always due to a high accident rate. Witnesses recorded in the 1864 report told of club money being appropriated by the adventurers. One described three or four hundred pounds being taken out to pay merchants' bills. In another case, when a mine closed before the surgeon was even appointed, the adventurers used the cash to treat themselves to a grand dinner. One mine captain, W. Rutter, on finding that the club money had all been used to help restart the mine, expressed his utter disgust: 'I told the purser, then and there, tooth and nail, that they were taking the blood and bones of the men.' In spite of such wretched behaviour, many of the clubs were properly run and, in a few cases, the miners could make an extra payment to obtain medical help for the whole family. But the fact remained that in the great majority of cases, the adventurers ran the clubs and the men who paid the money had no say in its management.

At the end of the miner's life, the funeral was an important event for the whole community. With the spread of Methodism, the solemnity was lightened by a special funeral song that accompanied the party as they moved from the chapel to the grave:

'Sing from the chamber to the grave'
I hear the dying miner say
'A sound of melody I crave
Upon my burial day'.

'Sing sweetly whilst you travel on
And keep the funeral slow;
The angels sing where I am gone
And you sing below.

'Then bear me gently to my grave,
An as you pass along,
Remember 'twas my wish to have
A pleasant funeral song.'

It might seem that the miner's life was one of unremitting hardship and gloom, but there were a few days a year when all work stopped. They were indeed few and far between. Christmas and Good Friday were everywhere observed and in some districts there were half-days at Midsummer and Whitsun. The best days of all were the parish feasts, whole weekends given up to pleasure, with traditional games, such as Cornish wrestling and quoits and a great deal of drinking before Methodism became so widespread, anything from beer from the alehouses to illicit home brewed spirits. Inevitably, those in authorities frowned on these activities but they were bright spots in dull existences, a relief from the endless labour that began in childhood.

Chapter 9

CHILDREN AT WORK

The most comprehensive survey of the lives of children working in the mines was contained in a Report published in 1842. A year earlier, Lord Ashley, later Lord Shaftsbury, had commissioned a report into the work of children in all the mining districts of Britain. The Commissioner appointed to investigate the South West was Charles Barnham, a doctor who had considerable experience as a mine surgeon and who had made a special study of diseases peculiar to workers in the tin and copper mines. He was, therefore, the ideal person for the job. He interviewed extensively and the scale of the task facing him can be gauged by the numbers of children he recorded as working in the industry in Cornwall and Devon: 15,342 boys and 9,957 girls. His Report, published in 1842, does not make comfortable reading.

Children were employed both on the surface and underground, but there are no recorded cases of girls ever working down the mines. The main job for the surface workers was dressing the ore and the hours varied, depending on a number of factors. The most obvious limit was daylight, so in summer the working day generally lasted for ten hours, starting at seven in the morning, and in winter nine hours, from dawn to dusk. There was always a midday break for a meal that might be as little as half an hour at some mines, but could be a whole hour and in a very small number of mines two hours were allowed. These were the normal working days, but life became more hectic when the ore had to be prepared for sampling ready to be sold. The intervals between samplings could vary from two a month to once every two months. If extra hands were employed, then the working day remained much the same as usual, but if no one else was taken on, the children faced extraordinarily long hours.

> 'A boy or girl, from nine to twelve years old, is obliged to arise at about four o'clock in the morning, gets a hasty breakfast, and after a walk of half an hour or more – three or four miles – reaches the mine at six. Work is continued till twelve, without intermission or refreshment, save what may be got by stealth. Half an hour is then employed in taking dinner. The child

> then works without interruption till eight; gets home, after repeating the walk of the morning; and may have had supper and gets to bed about ten. It is chiefly the younger children who are called upon to begin their work at six o'clock, the process on which they are engaged being preparatory for the others.'

It makes for chilling reading, and one can only imagine the wretched children dragging their feet as they stumbled home, exhausted. The Report speaks of a half hour walk, but one would need to be a super fit athlete to walk three or four miles in such a short time. In a few cases, boys were employed on ore preparation at the busiest time, in shifts that started at seven one morning and only ended at two on the following afternoon. This exhausting work was, the Report blithely announces, welcomed by the boys because they got extra pocket money only for themselves. The Report mentioned that the long shifts went to the youngest and a table shows the ages for children doing surface work, and although in the vast majority of cases few are employed under the age of seven, a mine at St. Ives recorded a five year old at work.

Boys working underground had exactly the same hours as the men they worked with. The work went on day and night, generally divided into three eight-hour shifts or 'courses', as they were known, starting at 6am, 2pm and 10pm. If the work was exceptionally hard, the day might be split into four six-hour sessions. The changeover was not made at the surface, but at the foot of the shaft or, more usually, at the working face, so the boys would have to add on the time taken in climbing the ladders to their actual work underground and walking from the foot of the shaft to the face being worked. The men usually arranged to rotate the courses, so that a team who were working a night shift one week would be on a day shift the next and so on.

In deep, hot mines, few could manage to work more than the standard eight hours, but when men came across a rich lode, they would want to extract as much ore as they could at the price being paid at the time, before the adventurers could change it. On other occasions they would work the extra hours simply to make up for a colleague who was away sick or injured. They would then work what was known as 'double stem' – or, in other words, double shifts. Some of the most robust miners would work almost non-stop for up to sixteen hours. If the men worked double stem, then so too did the boys. On other occasions, they would work the extra simply to make up for a colleague who was away sick or injured. One miner interviewed by the Commission, Thomas Moyle, reported that as a boy he had once worked five double stems, one after the other. Once again, as with the surface workers, the boys were said not to mind the extra hours because they got pocket money to spend on themselves, rather than having to hand it over as part of the family income. Boys started work below ground at a later age than surface workers. In a list of sixteen mines, there were only three where boys under ten went below ground – two at age nine and one at seven. The rest were all aged between eleven and thirteen when they started work. If the men were often exhausted, life for the boys was no easier, and in many cases where they were doing the hard work of wheeling ore and waste, they were released after six hours and allowed to go back to the surface ahead of time.

The surface work depended on whether the mine was mainly producing copper or tin. Good stream tin needed little preparation and could go straight to the smelter. Dealing with copper ore was much more complex. In general, the valuable ore only represented some 9 per cent of the material mined. The first step involved separating out the small and large pieces of rock, before being sent to the 'pickers', who were mostly girls. Young boys were often involved in the first part of the process, which was washing the stones in a trough of running water. This must have been a miserable occupation on a cold winter day, and because the washing was done using a sieve, the children often complained of back ache as well as the cold. The water slopping around meant that their feet were often soaking wet soon after they started work.

The youngest girls were the 'pickers' who had the job of separating out the valuable material from the waste. They got the job simply because it required the least physical strength of any of the various operations. That does not mean that it was either easy or comfortable. The little girls sat at long tables. Boys would bring the stones and heap them in front of the girls, who would then begin the task of picking. Copper ore is generally brightly coloured, usually with a greenish-blue shade, while tin ore is silvery. Both stand out against the granite. Good material was thrown into baskets in front of the table and the waste was thrown into 'boxes' to be wheeled away by boys and added to a large heap, which would then be looked at again to make sure no valuable ore had been missed. The work station had a roof to keep off rain and snow, but was open at front and back to allow the boys to bring in stone and take out waste. One only has to look at the situation of so many engine houses, perched high on moorland or above sea cliffs to imagine just how unpleasant it must have been in bad weather with the wind howling through the open shed and driving in the rain. Cold was the greatest affliction these young children had to endure. Their only relief was the short lunch break. In winter, they would sometimes be fortunate enough to have a room set aside with some form of heating, but in many mines the best they could do was to huddle together in the 'Dries', the rooms where the men changed their clothing. Summertime was far more pleasant, when they could sit out on the grass. Boys and girls seldom mixed, with the girls taking their time and the boys bolting their food to have time for a few games. The lives of these young children were hard, though the actual work was considered light. For boys going underground, things were even tougher, but they were usually older starting work than the surface workers. Out of a sample of fifteen mines, there boys under eleven, though one reported a boy of just seven and a half.

The young boys often started by working the 'air pump'. When mines go down to great depths, the air is often very bad and the machine is devised to draw in the foul air and exhaust it at a higher level. It is very basic, consisting of a box, filled with water, inside which an inverted box is moved up and down by a simple pump handle. The work was described as being very light, though perhaps few would relish the idea of spending several hours a day working a pump handle up and down. And it was also usually the case that the area where the boys worked contained some of the poorest quality air, which often brought on headaches and muscle pains. Another common task was known as 'hauling tackle', bringing up spoil during the sinking of shafts, either to the surface or between different underground levels. This was done using a simple, hand-operated windlass that hauled up the heavy iron bucket full of material. The work was strenuous, but the boys involved at least had purer air to breathe than those on the air pumps, since they generally worked at higher levels.

The most common work carried out by the boys was called 'rulling' or rolling, which was wheeling material from the face where the men were working at the end of the level to the foot of the shaft. There the barrows were emptied into a large iron vessel, a kibble, and raised to the surface. This was extremely hard work, as the face and the shaft could be as much as 200 yards apart, and the ground extremely bumpy. The boys would be kept going almost continuously – one filling the barrow, another wheeling it. The area where the barrows were filled being right at the end of the level, it was the least well ventilated and the boys suffered from what they called 'bad air'. Charles Manuel began this work at the age of thirteen, but soon found himself regularly spitting out 'black stuff'. He soon began to suffer from coughing fits, and his health became so bad that eventually he had to take a job back at the surface. He was not the only one who suffered in the same way. It did not have to be like that.

In 1776, a mining engineer, John Curr, developed a system of laying a simple track of iron rails or plates to link the working area to the shaft, on which wheeled wagons known as trams could be run. Moving these was far easier than wheeling barrows, and an anonymous poet on Tyneside wrote this poem to thank Curr – it should be read with a Geordie accent.

'God bless the man wi' peace and plenty
That first invented metal plates
Draw out his years to five time twenty
Then slide him through the heavenly gates.

For if th human frame to spare
Frae toil an' pain ayont conceivin'
Has aught te de wi' getting' there,
Aw think he mun gan' strite to heaven.'

Where the tin and copper mines adopted the system, the work became much less strenuous. Boys who were 'tramming' reported that they found the work comparatively easy. Unfortunately, in many mines the old ways persisted and too many boys suffered real hardship. James Steven went rolling at a young age and, although he never worked nights, his life was one of miserable poverty. He would get out of bed at four in the morning and seldom had time to have breakfast, but grabbed a hunk of bread and butter to eat on his way to work. His next meal was when he got home and that was frugal. By seven, he was in bed. There was only one reason why such conditions had to be endured – poverty. This was a family of nine and four of the children too young for work had to be fed. Necessity forced parents to take hard decisions. Henry Trevethan was fifty when he was interviewed in 1841 and he told how, twenty-six years earlier, he had lost one eye in an explosion, but was back at work within three weeks. Then he began to suffer the familiar complaints arising from 'bad air' and had to leave the better paid underground work and get surface work. He had no option other than to send his own children to work at an early age and was so poor he could not afford the few pennies needed for even the most basic education. The parents were not cruel and heartless but lived in a system that all too often was. Some companies were more concerned for their workers' welfare than others. At the great Dolcoath mine, the management set up warm rooms for the men when they came up from their stint below ground and gave each of them a bowl of hot soup. This was said to cost the company no more than forty shillings a month and was reported to have had a great effect in improving the health of the men and boys.

The final stage reached by the boys came when they started to help with boring, creating the holes in the rock for gunpowder and blasting. The youngest boys often assisted in the early stage, kneading the clay to a steady consistency that was then tamped into the hole after the powder was in place, to seal in the blast. The older boys would start by holding the iron drills, turning them at each blow of the hammer and, as they got more experienced and stronger, took their turn at hammering in the spiked drill. At this stage, they were usually 'taken into concern' being treated as 'half men' and receiving their own share of what the men were paid. This was a matter of great pride, but also of changed attitudes. They were no longer treated as boys but as men and expected to do the same work as all the other adults. Boring was not only a strenuous job, but a dangerous one. Accidents were frequent and it was difficult to keep a steady rhythm when hammering; it only needed a badly aimed blow to cause a severe strain. And because they were always working at the dead end of the level, the fine dust that flew up at each blow was never dissipated but hung in the air as an invisible but deadly cloud.

It is difficult to find out how many children died in accidents as ages were seldom recorded, but they did occur both above and below ground. Accidents at the ore crushing mill were rare and when they did occur were dismissed as due to the 'heedlessness natural to boys'. In general, it seems the fault always lay with the children. At Wheal Vor, two boys were told to remove a great heap of wet soil. They seem to have decided the task would be easier if instead of working from the top, they shovelled away at the bottom to spread it out first. Tragically, it collapsed on top of them, killing one and seriously injuring the other. But these accidents were rare compared with those underground.

Most accidents in the mines occurred during blasting and, as only the older boys were involved, the younger ones were not at risk. But the younger ones did have a special danger of their own. The 1842 Report mentions injuries caused by falling off ladders, simply because the very young simply lost the strength they needed to hold on, which is a shocking indictment of the system that put them in that situation. The other major cause of fatalities was falling down pits, particularly while wheeling barrows. Inevitably, these accidents were generally attributed to carelessness, but the Report did mention that candles did sometimes go out in foul air, leaving the boys stumbling around dangerous places in the dark. It is all very gloomy, and the following extract from Wilkie Collins' *Rambles Beyond Railways* (1851) provides an entertaining change of mood. He was a short man, just five foot six and he was kitted out in clothes intended for a six foot two miner, which made getting around difficult, even when sleeves and trousers were rolled up. Then they came to a place with widely spaced beams over a pit that he was quite unable to manage in his oversized outfit.

> 'Our friend the miner saw my difficulties, of extricating me from it at once, with a promptitude and skill which deserve record. Descending halfway by the beams, he clutched with one hand that hinder part of my too voluminous nether garments, which presented the broadest superfices of canvas to his grasp (I hope the delicate reader appreciates my ingenious indirectness of expression, when I touch on the unmentionable subject of trousers!). Grappling me thus, and supporting himself by his free hand, he lifted me up as easily as if I had been a small parcel; then carried me horizontally along the loose boards, like a refractory little boy borne off by the usher to the master's birch; or – considering the candle burning on my hat, and the necessity of elevating my position by as lofty a comparison as I can make – like a flying Mercury with a star on his head: and finally deposited me safely upon my legs again, on the firm rock pathway beyond.'

It is humorous, but if one thinks back to the accident reports, it takes on a more sinister aspect. One can all too easily imagine a small boy coming on his own to such a spot and trying to get himself across without slipping or stumbling.

One authority did give information on children's mortality. Sir Charles Lenson in an article for the *Journal of the Statistical Society of London, 1838,* quoted figures on children's mortality in copper mines in three districts. In Gwennap in 18 months up to January 1837, three children, between the ages of 10 and 20, died in mine accidents; for Redruth the deaths over seven years were seven and for Illogen, five in five years. What is more startling is to find that compared with these fifteen accidental deaths, sixteen young people died from 'Diseases of chest'.

Many accidents now seem avoidable, and one of the recommendations made in the final Report of 1842 was that no boys should be allowed to work underground until they were fifteen years old, which would mean that though the use of ladders was still pernicious, at least they would be strong enough to climb them. It also recommended that no one should be employed in any part of the mines under the age of nine. For surface workers, no child should work more than eight hours in any twenty-four hour period until they were thirteen, when the maximum should be ten hours. There was also a proposal that at least three quarters of an hour should be allowed for the midday meal.

The one area that was touched only briefly was the pay. For surface workers this was very straightforward as they were paid by the company. The parents would bring a child along and as soon as an opening was available they would start to work at the rate of a penny or tuppence a day. As they got older, the rate generally went up, but this was still very low with an average of around five shillings a week. Usually, one older child was deputed to go to the count house for the monthly payment for a group of half a dozen. In the better establishments, they were paid in coins that could easily be divided up, otherwise they had to go to a shop or ale house for change. The children handed over the money to the parents, but when they got special payments for extra work they usually were able to keep the money for themselves. It was a rare treat to have money of their own to spend on whatever took their fancy.

The system for underground working boys was more complex. Where they were paid directly by the company, it was uncomplicated. Like the surface workers they were paid at the count house. Again, payments varied with age and experience. At the Levant mine, for example, the youngest worker at nine got 2s 6d a month while a seventeen year old got 13s 6d. But these were in a minority. Most boys were employed by the older men. They too were supposed to be paid monthly, but the men were only paid every two months. By the end of the period, the money in the family often ran out, and the men had to borrow on 'subsist' until the next pay day. As a result, there was not enough to pay the boys so the manager would step in. The boys got their money, but the family debt went up. As with surface workers, basic pay went to the parents, money from extra work was kept by the boys. Altogether the picture that is painted is of poverty, with families always trying to catch up, but somehow never quite managing to survive to the next pay day. It is all in all a sorry tale of hard work, long hours and low pay.

The Report was sent to parliament, along with others from other regions, mainly coal mining areas. The result was the Mines and Collieries Act of 1842. Its recommendations did not go as far as Barnham had recommended in his report and had little effect on conditions in the South West. The main changes to the law made it illegal for girls to work underground, which never happened in the tin and copper mines anyway, and specified that no boys should go underground until they were ten years old – a long way short of the fifteen years suggested by Dr Barnham. The Act made huge differences to the lives of children in collieries but had little effect in the mines of Devon and Cornwall.

Chapter 10

SMELTERS AND FOUNDRIES

As mentioned before, tin smelting was comparatively straightforward in comparison with copper smelting, nevertheless there were numerous attempts, mainly unsuccessful, to establish copper smelters in Cornwall. The major breakthrough came when an experienced metallurgist obtained the backing of local men of influence, such as John Vivian and Sir John St. Aubyn, to establish a smelter near Camborne. The trouble was that although it was conveniently close to the major copper mines, where ore was readily available, it was a long way from the north coast. One authority quoted figures to show that as much as eighteen tons of coal was needed to smelt one ton of ore, and the nearest source of coal was South Wales. The cost of bringing coal from the port to the smelter was eating away the profits. Just two years after the smelter began work it was decided to move the whole operation to the coast, a decision that was to lead to a new industrial town being established at Hayle.

The name 'Hayle' is simply the Cornish for estuary, and that was the attraction. The site was on St. Ives Bay and was chosen because there are two tidal lakes that form a natural, protected harbour. The larger of the two to the west is fed by the River Hayle, the smaller by the rivers Angarrack and Penpol. The only settlement at the time was the village of Phillack, now chiefly known for two very different buildings, the ancient church and the Bucket of Blood pub, named after a gruesome murder. The smelting business, now known as the Cornish Copper Company, was established to the east of the estuary, and the smaller lake soon had a new name – the Copperhouse Pool.

Originally, the quays were on the west bank of the Hayle. There was always a problem getting in and out of the harbour because of a sand bar at the mouth of the estuary, where Lelant was the main port. But in 1740, a company was formed to develop port facilities to the east. By this date, the Newcomen engines had become widespread and their voracious appetite for coal created the demand for trade with Wales. Other mine materials, such as timber for lining shafts and for props were brought in, while ore could be exported. It was to the east that the Copper Company settled and built wharves along the Angarrack River. Now the smelters could be put to work. It was a complex process, involving the use of twenty-five reverberatory furnaces, that is, furnaces where the ore and the fuel never make direct contact, and the ore is heated by the hot gases.

The first stage was to calcine the ore, which drove off the arsenic in the mineral, which escaped as fumes. As there was little if anything to protect the men working on this stage, the risks to health are all too obvious. Four tons of ore were treated at a time and heated for twenty-four hours. Four hundredweight of the calcined ore was fed into the smelter and after four hours, the slag was run off and a second charge added; the process was repeated, until a total change of ten hundredweight had been dealt with. At the end of that process, the molten metal was run off into water to break it up. Then the whole process started up again. The first run off was calcined again, smelted again and run off. After that there was a third session of calcining and smelting before the molten metal was finally run off into moulds to be cast into bars. There were several different types of metal – each with a different use. Shot-copper for example was used for making brass, usually shipped out to Bristol and the brass mills along the River Avon. Some, mixed with tin, was sent as bell metal. Copper was also sent a short distance to St. Erth, where the ingots were either treated by tilt hammers or passed through the rolling mills to create copper sheets. Some 400 tons of copper were treated each year at the plant at one time.

An opportunity arose in the 1780s to improve communications. Merchant Curnow, one of the partnership who had first established the quays in the area, died and the Copper Company were able to buy up the concern, acquiring two extra quays and the trading business. They at once set about an improvement scheme, dredging and widening the Angarrack River to create a canal, with a new wharf constructed from blocks of slag. The accumulation of sand was always a problem, but that was partially solved by constructing flood gates to impound the water, which could then be released in a flood to scour the channel. The smelting company were now successful traders. By the time work was completed, vessels of up to 120 tons could reach the works themselves, while ships of 250 tons could reach the coal wharves at the seaward end of the canalised river. Although smelting in Cornwall was never really viable, so that in time the vast majority of ore was sent to Swansea, the company's trading business was thriving. Then, to their horror, a competitor appeared, right on their doorstep – John Harvey.

Harvey began his working life as a blacksmith with a forge at Carnhell Green, a village some three miles from Hayle. He was clearly both a skilled craftsman and an ambitious individual. He was well aware that the world of mining was changing rapidly and that the new generation of steam engines were pumping water from ever increasing depths. In the early days, the pipes for pump work were crude, little more than hollowed out tree trunks. Casting pipes from iron was not straightforward and they were being turned out at the famous Darby works at Coalbrookdale in large quantities. However, it was a long way from Shropshire to Cornwall and there was clearly a market for any that could be made locally. But a village blacksmith rarely has the capital to set up his own foundry. The story of how Harvey got started might be one of the many myths that accompany engineering advances, but there is a core of truth in it, for he did get finance.

The story goes that Sir John St. Aubyn was on his way to church after a Sunday morning hunting, when he discovered that he had lost a silver buckle from one of his shoes. As he was close to the Harvey forge, he asked if the smith could help. Harvey fetched a silver spoon and contrived to shape it into a very respectable buckle. St. Aubyn was so impressed he offered to finance Harvey if he had any new ventures in mind. It is a good but unlikely tale – gentlemen do not normally go hunting in silver buckled shoes, nor do village blacksmiths have silver spoons tucked away in the kitchen drawer. But it does seem likely that Harvey's reputation was so high locally that he could attract finance from a rich gentleman who had already shown his willingness to invest in local industry.

In 1779, work began on building the foundry at Carnsew between the Hayle and Penpol rivers. Once established, Harvey was able to develop the necessary skills for casting pipes and the investment began to pay handsome dividends. There followed ten years of increasing prosperity. The arrival of Boulton and Watt engines increased the demand for iron castings and the only thing in the way of an expanding business was transport. This was monopolised by the Copper Company and Harvey was only one of their customers; he now found it increasingly difficult to get the supplies of iron and coal when he needed them. For Harvey, the solution seemed obvious and he bought a sloop, *Providence,* to control his own supply lines. The Cornish Copper Company saw the situation very differently. To them it appeared that a new trader had appeared to challenge their profitable monopoly. They decided to do everything they could to hinder Harvey. They bought up land near Carnsew, claimed rights to the foreshore and left Harvey with only one approach, down the River Penpol. This was nowhere nearly as accessible as the Angarrack and could only be used at very high tides. It was the first shot in a battle that would rage between the two companies for decades.

A village blacksmith who has by his own efforts become an important industrialist was not going to be easily intimidated. Harvey responded by chartering more vessels and doing just what the Copper Company had tried to prevent; he began setting up as a trader in all kinds of commodities from bricks to soap, as well as the staples of iron, coal and timber. He chartered vessels and he ordered a brig, a two-masted square rigger, from Bideford, which he named *Henry* after his son, who had now reached an age where he could take an active part in the business. Vessels could unload on the sand bar, which at low tide could be reached by carts and pack animals for unloading and carrying the cargo into Hayle. The Copperhouse foundry actually tried to sabotage operations by throwing cargo from the lighter overboard. When that failed to deter Harvey, they took more subversive action. One of the key elements at the foundry was the boring mill, used for such tasks as making accurately proportioned cylinders. The mill was powered by a water wheel, fed by a leat, and they were able to arrange to have the outflow blocked, effectively bringing work at the mill to a halt. But none of their efforts prevented the Harvey firm from growing and prospering. Nevertheless, the war was continued through several lawsuits, which cost both sides money and achieved nothing apart from making lawyers rich.

When John Harvey died in 1803, Henry took over the business and continued to expand both the trading and foundry sides. He built up the fleet of sailing ships and by 1831 he had moved to steam, running a regular steam packet service between Hayle and Bristol; the Great Western Railway did not reach Cornwall until 1859, with the opening of the Tamar bridge. The company was so successful that businessmen were arriving in Hayle from all over the country. There was no appropriate accommodation in this new industrial town, so Harvey built his own hotel, the White Hart, to entertain them. In 1816, Richard Trevithick set sail for Peru to oversee the installation and running of his high pressure steam engine at a silver mine in the Andes. He was to be gone for eleven years and in the meantime his wife Jane, Henry Harvey's sister, was helped out financially by being put in charge of the new hotel. The connection between the Harvey family and Trevithick was not just one of family. When the Boulton and Watt patent expired, there was a flurry of activity among Cornish engineers. One of these was Richard Trevithick, who was now married to Harvey's daughter Jane. In 1801 the Harvey foundry supplied parts for his new invention, a steam locomotive. This was intended for use on ordinary roads and received its first success by roaring up Camborne Hill on Christmas Eve of that year. Trevithick would soon go on to place his locomotive on a railed track and inaugurate the age of the railway.

Meanwhile, the Copperhouse company had finally abandoned the unprofitable business of smelting copper and started to manufacture steam engines. Their first customers were the local mines, but in time they expanded their activities. A new customer appeared as the demand for fresh water grew and large waterworks were established. One of these in London was the Grand Junction and Copperhouse foundry, now known as Sandys, Carne & Vivian, built them a massive engine with a 90 inch diameter cylinder. The machine initially drove a 33 inch diameter plunger pump that delivered 5.5 million gallons of water a day. It is now preserved at the London Museum of Water and Steam at Kew and claims to be the largest working beam engine in the world. By this time, the foundry was employing around 300 men

Harvey was not slow to see that engine building was a potentially lucrative business. Harvey's of Hayle was to become famous as a builder of magnificent steam engines. The company began modestly in 1830 with a 14-inch whim engine for Wheal Towan and over the next decade built a total of 82 engines and not just serving local customers. In 1835, a Harvey engine was being carried across the Atlantic to Virginia. The biggest job ever undertaken by the company was building huge engines to drain the Haarlem Mere in the Netherlands. The first cylinder was cast in 1843 and was 144 inches in diameter, requiring 25 tons of metal. It was the first of three similar engines to be built and one of these, the Cruquis engine, is preserved as a museum. The huge engines were required for a massive task, draining a thousand million tons of water from an area of seventy square miles. It was an immense achievement and was one of the factors in establishing the Harvey reputation and bringing the company orders from as far away as Australia. They built their last engine in 1891. Matthew Boulton had boasted he would build engines for the whole world; Harveys did. It was a tribute to the enterprising spirit of the Cornish engineers that, once free from restrictive patents, they could produce some of the finest stationary steam engines ever built. The great rivalry came to an end in 1867, when Copperhouse went out of business and in the ultimate irony was bought up by Harvey.

Harvey and Copperhouse were not the only engine manufacturers in Cornwall. A group of wealthy individuals got together to finance a foundry at Perran Wharf on an inlet of the River Fal in 1791. They were the first to grasp the opportunity provided by the loss of the Boulton and Watt patent. Among the chief investors were the Fox family of Falmouth, who already had considerable connections with industry in South Wales, owning collieries and iron mines. This was to prove invaluable. The new concern leased the extensive Neath Abbey ironworks in South Wales and now had access to everything they needed. Ore from the Fox mines could be smelted at Neath Abbey and sent out as pig iron to join the coal from the collieries. The material was brought to Portreath, north west of Redruth. Visiting the village today, with its narrow harbour, it is difficult to imagine what it would have been like in those days, when it was a bustling port and in the early nineteenth century regarded as the most important in the county. There was also considerable trade down the River Fal to the foundry and, although the creek was too shallow to accept sea going vessels, it could be used by the company's barges who took engine parts down to Restronguet Creek wharf, where a large crane was available to load them onto ships. The foundry prospered and at its height occupied a site covering 6 acres and employing around 400 men. It had as extensive an export trade as Harvey's, supplying engines to Europe and South America.

Among the other foundries, one of the most important was Holman Brothers, started in 1801 at Pool on the outskirts of Camborne. The company was highly successful, moving on from manufacturing steam engines to producing new types of mining equipment. In 1968, Holman's merged with another company to form Compair Holman. Then, after two centuries the name Holman disappeared from the list of Cornish manufacturers, when production moved to Germany, The Charlestown Foundry was built by the harbour at Charlestown. The Charles in question was Charles Rashleigh, who built the harbour to serve the china clay works at St. Austell. The foundry was very handy for shipping, but not so convenient for sending engines out to the mining areas. This was always a dilemma when siting foundries; if one was near the sea for shipping in coal and iron, one was not necessarily well placed for sending the engine parts for erection at the mine. That in itself was a major undertaking.

In 1843, William Francis published a long poem, describing in detail the work involved in erecting an engine. The engine house would have to be ready to receive it, and constructing that was in itself a major task. The bob wall that carried the whole weight of the beam had to be robust enough and granite was usually specified. Nicholls Williams & Co were building an engine house near Calstock, and recorded having problems:

> 'We opened a quarry near the old shaft for the purpose of raising stone but finding it not sufficiently strong I thought it advisable to go to Hingston

Down for the stone to build the engine-house where we have a large supply of stone sufficiently strong, some of which are half a ton weight.'

The quarries there obviously contain good quality stone as they are still being worked. The engine house complete, it was time to install the engine and one of the first jobs was to set in place the iron plates on the short 'bob wall' to provide the pivot point for the overhead beam. The preparatory work involved erecting shear legs, with block and tackle and a capstan which would be manned by several men to do the lifting. This is the verse description of that part of the operation:

'Then comes the main object, the main beam to hoist,
And over the shaft thirty tons equipoise,
And lift to its plate, this the capstan achieves,
While winches and blocks the great pressure relieves.'

This was the most difficult part of the whole operation and there were anxious moments until the beam was secured firmly on the bob wall. Thirty tons is a huge weight to lift manually, even with a great deal of tackle, but beams could be even heavier. The preserved 80 inch engine over the Taylor shaft at the East Pool mine is 33 feet long and weighs in at 52 tons. Altogether the iron work in the engine weighs 125 tons, but if you add in all the necessary pumping equipment, 684 tons of ironwork would have been brought to the site.

Given the huge effort involved in building and preparing an engine and its engine house, it was only right and proper to celebrate the event by holding an official naming day with appropriate celebrations. The church bells would ring, the engine would be garlanded, a band would play and speeches would be made before the actual naming ceremony took place. A bottle of champagne would be broken over the beam as the engine began its slow bobbing. The engineer would be congratulated on setting it in motion so smoothly on the first go, though everyone knew that there would have been several unofficial trials before the big day to sort out any problems. The naming complete, everyone would go off for a feast; the adventurers to wine and a banquet in a marquee; the miners to beer and modest food out in the open. Big engines got even grander treatment. At the naming of the hundred inch engine at Wheal Vor, there was a fireworks display and three brass bands to entertain the crowd. This was the culmination of a long process that had begun with the construction of the engine house and the bringing of vast quantities of material, often over long distances, from the foundry to the mine. In the early days of engine construction, roads were generally poor and the final approach to a mine site was usually over a rough track. Engineers looked to other forms of transport.

In the second half of the eighteenth century, Britain as a whole saw a huge programme of canal construction getting under way, but the nature of the landscape in Devon and Cornwall limited the areas where canals could be built. One successful venture already mentioned was the Tavistock Canal, linking mines to the port at Morwellham. This was followed by a canal linking Liskeard to the sea at Looe, which served the mines and quarries of the Cheesewring. Little of this waterway survives, for it lost most of its trade when the railway was built on a very similar route. But canals only ever played a minor role in transport. The mineral railways were far more significant.

The mines at Gwennap were particularly badly placed for transport, being right in the middle of the Cornish peninsula. Local mine owners set up a subscription in the 1790s for a survey of a possible canal to the north coast. Francis Basset, an important landowner, recorded contributing ten guineas for John Williams of Scorrier to carry out a survey, but nothing came of it. Certainly, it is difficult to see how the canal could have been supplied with water. There was an alternative. The collieries of north east England and the mining and industrial areas of South Wales were already covered with an intricate network of tramways or plateways. These were early rail systems along which wagons were pulled by horses. Unlike the edge rails we see on modern railways to carry trains with flanged wheels, these rails were L-shaped in cross section and the wagons had plain wheels, kept on track by the upright parts of the plates. Also, because the trucks were pulled by horses, the space between the rails had to be kept clear. So, instead of the transverse sleepers we see today, the rails were mounted on rows of stone blocks. A hole was drilled in the centre of each block, then filled with a wooden plug into which an iron spike could be driven to hold the rail in place. Many of these sleeper blocks have survived, long after the lines were abandoned and the rails removed. Having given up on the notion of canal construction, John Williams, with the backing of the Fox family, now laid out what would become Cornwall's first railway – the Portreath Plateway – in 1809, linking the mines to the port. John Taylor, who had been engineer for the Tavistock Canal, then set about building a tramway from his mines at Gwennap, a necessity if he was to compete with the Williams interests. The Redruth and Chacewater tramway ran from Gwennap to the newly developed port of Devoran on the Fal.

Joseph Teffrey was another mine owner who considered canal building. In 1815, he drew up plans for a canal that would link the Fowey Consols copper mines to Padstow. This came to nothing but, in 1829, he did construct a canal from Par on the south coast to the foot of Penpillick Hill, later extended to Port Mill. Here an inclined plane was constructed; this was a simple railed track up the hillside, along which trucks could be hauled by water-powered machinery. It connected with a tramway leading to the mines. The next stage of development gave Cornwall one of its most imposing industrial monuments. The tramway was extended to a spot near Indian Queens. It required another water-powered incline, but it also had to run across the deep Luxulyan Valley. A granite viaduct was constructed, 660 feet long, carried on ten arches and rising to almost a hundred feet above the valley floor. It also served as an aqueduct to carry water to power machinery in the bottom of the valley.

The tramways did their work well but, in 1832, work began on the Bodmin and Wadebridge Railway, which was not to be worked by horses but by steam locomotives. This was soon followed by the first mineral line to be built for locomotives. Harvey's decided that the time was ripe to construct a line. Their railway headed right to the heart of one of the most prosperous mining districts around Camborne and then branched off to Redruth, with branches to Tresavean and Portreath. Steam locomotives were in use from the start, and it is appropriate that one of the first railways in Cornwall to be worked by steam had helped Richard Trevithick to build his first, pioneering engine. Harvey's, as always, were quick to see the commercial possibilities the line offered, so on Sundays, when there was no goods traffic, they ran excursion trains from Camborne and Redruth to the seaside at Hayle. Rev Thomas Collins was one vicar who ran Sunday school trips and lauded the excursions in verse.

'We shall see the rolling ocean,
We shall breathe the fresh sea air,
See, the country, comes to greet us,
And the swallows can't outfly.
Houses, trees and edges greet us,
Running by, and running by.
Happy Camborne!
Where the railway is so nigh.

One other railway deserves a special mention. It was begun in 1863 as the Tamar, Kit Hill and Callington Railway but was soon renamed simply as the East Cornwall Mineral Railway. It linked the extensive mines in the Kit Hill area with the Tamar quays at Calstock. The line, worked by a pair of locomotives, arrived at Calstock high above the river, and was joined to the quay by an incline. What makes it especially interesting is what happened when the line was taken over in 1901 by the Plymouth, Devonport and Great Western Junction Railway. It is a good rule of thumb when looking at railway history to assume that a company's importance is in inverse proportion to the grandeur of its name. This was never more than a modest line, but it did and does boast one magnificent structure, the Calstock viaduct. It would be easy to mistake this for one of the many splendid stone viaducts built by Victorian engineers, but this twentieth century masterpiece, 1,000 feet long and standing 117 feet above the Tamar, is actually constructed out of more than 10,000 concrete blocks. It also had a unique feature; the old incline was abandoned and a vertical lift ran up one of the piers to carry wagons up and down between the tracks and the quay.

The story of the development of the foundries that built the great engines for the mines is a vital part not only of the whole story of mining in the region, but with the development of so much more, from the creation of harbours to the building of a rail network. These were works that were not just vital to the development of the mines but were to prove of value to the whole community.

Chapter 11

THE WORKING MEN

In 1864, a Parliamentary Commission was set up to enquire into conditions in the British mines but excluding collieries. The final Report provided a fund of information on conditions, as well as quoting extensively from interviews with individual miners. It gives a unique insight into the working life of the men in the middle of the nineteenth century.

One of the more obvious findings was that the earnings of miners were uncertain, varying enormously from time to time, depending not just on the state of the industry but also on the difficulty of the work undertaken. And there was one patent injustice. Men on wages were paid for four weeks' work, which was known as a month. But they were only paid at the end of the calendar month. February is the only month of the year with 28 days in it, so for example a man working every day in August would work 31 days, three of which would be unpaid.

The same system still applied, with the surface workers on wages and the miners divided between tutworkers and tributers. Although the three branches all made similar amounts of money when looked at over a long period, the tributers were generally slightly better off and were certainly regarded as the elite among the mining community. Both tut and tribute carried an element of risk. A mine captain described what could happen to a group of tutworkers who had taken on a contract at a fixed rate: 'You get a pare of men and give them £4 a fathom and before they have driven six feet the ground will become such that they ought to have £7 or £8 to make a living.' Needless to say, there was never any question of that extra money being paid. There was even greater risk when it came to the tributers, who if they hit a good patch could do very well, but could also find themselves in real financial difficulties when things went wrong. As described earlier, the miners often had to borrow on subsist, and when they had paid that back, together with all the costs they had incurred, could be left with little or nothing, and had to go back again on subsist to survive until the next pay day. The tributers always lived in hope of hitting a really profitable lode and making a great deal of money. But they also dreaded those times when everything seemed to go wrong. Captain Joseph Vivian described a case of the latter to the 1864 Commissioners.

> 'We have a fine-rate old man who has been working in North Roskear for thirty years, the last two years he has always been speculating and has done badly. I said to him the other day, "You are doing badly?" "Yes," said he. "I never had such a long run before, but I shall make it up again soon."
>
> 'On the whole he has not done badly. He is worth £200 or £300 probably. He has one or two cottages and keeps a cow, and so on. We never let him go upon less than £2 5s a month, though for a long time he has not earned any money.'

He was fortunate in having an understanding company who were prepared to help out, and from the description of his possessions, he was comparatively well off. £200 would be worth nearly £20,000 today. Few tributers could have saved that much, which perhaps explains why he did not seem too concerned for the future. Others were not so fortunate. David Buzza, a tributer at Caradon, told a very different story.

> 'I worked in a mine called Devon Great Consols, and my family is now residing at Tavistock. I have quarters here. I have been turned away, after I had done very badly for some months. In fourteen months all that I earned was £15 1s 4d.'
>
> 'You gave up work there?'
>
> 'I complained I had not got money enough, and it was in their power to let me get more, and because I complained they turned me away out of that mine and would not let me work in any mine in the district. And this is how a miner is tossed about. I went to the different mines in the district and I could not get any more work. "I have no work for you", they said.'

One of the features of the mines of Cornwall and Devon was the absence of any sort of Trade Union until quite a late date. If a man such as Buzza received unfair treatment, there was no one else to argue his case. But the mine adventurers could decide he was a trouble maker and get together to have him banned from all the mines in the area. Bad times could mean not just loss of income, but also deteriorating health and strength from malnourishment.

At least one expert thought that the whole system should be changed. James Sims had been a mining engineer for 35 years when he wrote an article

The earliest representation of early mining technology can be found in Agricola's *De Re Metallica*, 1556. This illustration shows a series of simple pumps worked by a water wheel and crank. If the clothes of the miner look familiar it's because Walt Disney used these illustrations as the basis for the dwarfs in Snow White and the Seven Dwarfs.

The ore stamps at Blue Hills Tin Streams, worked by a water wheel. The projections on the rotating cylinders lift the vertical stamps that then fall back under gravity. The same technology has been in use for centuries. (Blue Hills Tin Streams)

The use of gunpowder for blasting was introduced to Britain from Germany. In this photograph from the end of the 19th century, a miner is tamping the charge home, after which the fuse will be lit. (Tony Brooks)

Cornish mines were notorious for the vertical ladders in the shafts, often descending for more than a thousand feet. These ladders are at an angle, and connected the 400 and 460 fathom levels of King Edward Mine and it is surprising to find such a system still in use at the beginning of the 20th century. (Tony Brooks)

A typical mine head scene in the 19th century; it shows a rotary beam engine at Cook's Kitchen mine.

This dramatic photograph gives a good impression of life at a stope, with heaps of spoil and the roof propped with a rough array of timbers. (Tony Brooks)

The man engine at Dolcoath mine. (Tony Brooks)

This derelict miner's cottage on the approach to Botallack mine, photographed in the 1970s, shows how the walls were built up of a random collection of stones collected from the surrounding countryside. (Clive Coote)

This grand house was originally home to a mine captain at the Duke of Bedford's mine in the Tamar Valley. Today, it is the gourmet hotel The Horn of Plenty. (Tony Brooks/ Horn of Plenty)

The proliferation of mines and engine houses round Camborne must regularly have left the residents under a cloud of smoke.

Children and bal maidens at work at the ore dressing sheds. The photograph clearly shows the dreadful conditions under which they had to work.

A small crane being used to empty a kibble into a wagon on an underground tramway. The introduction of railed tracks underground made the work of transporting ore to the foot of a shaft far less arduous.

Botallack mine a century ago. The Crown engine houses are now famous and people are amazed at the sight of the lower engine house on its tiny ledge, but, as the photograph shows, it once shared the constricted space with a boiler house.

Raising the headgear for a whim engine, using shear legs at King Edward Mine. (Tony Brooks)

A group of miners pose for the camera in the typical clothes worn in the nineteenth and early twentieth centuries. The candles on their hard hats are attached by a lump of clay. (Royal Institute of Cornwall)

The splendid engine house of the silver and lead mine Wheal Betsy on the edge of Dartmoor. (Nilfanion)

A familiar engine house, but not in England; this one stands above a copper mine at Allihies in County Cork, Ireland.

Another distinctively Cornish engine house can be seen behind the headgear, but this one is on the other side of the Atlantic at the Rio del Monte gold mine in Mexico.

Ned's stop at the Burra Burra mine in Australia. (Art Gallery of South Australia)

The sweep arm, cranks and flywheel of the Levant whim engine, with the cylinder in the background; the drive was taken to the winding drum outside the engine house. This is the last remaining beam engine in Cornwall still worked by steam. (The Trevithick Society)

A view of the 'Monster Mine' at Burra, from a book on South Australia by Jefferson Pickman Stow published in 1883.

The ore mill at King Edward Mine c.1914; the Frue vanner can be seen to the left of the picture. (Tony Brooks)

Women rarely went down mines, but this group of intrepid ladies are about to descend King Edward at the beginning of the twentieth century. (Tony Brooks)

By the end of the nineteenth century, horizontal steam engines were taking the place of the old beam engines for winding. This one was built by Holman's and originally installed at King Edward in 1908, but was later moved to Carn Brea where this photo was taken. It is now back at King Edward.

In its working days, Geevor was equipped with modern machinery, including this electric winding engine. (Nifanion)

The 32 foot diameter water wheel at Morwellham Quay that once powered machinery for crushing manganese.

in *The Mining Almanack for 1849*. In it, he questioned the value of the statistics regularly quoted that showed, for example, that on average men were earning 50 shillings a month, but gave no indication of how many were earning less, perhaps as little as ten shillings a month, out of which they might have to support a wife and family.

> 'Probably these men, when complaining to the agents of such small amount, and which has been no more for several months, are told, if you had worked as men ought, you would have got fair wages. But I would ask, what can be expected from a half-starved man, who is necessarily unable to do his work, and has the melancholy reflection, when at work, of seeing nothing but misery and starvation before him for himself and family. Not so with those men who are getting so much above the average as others are below the average. They, although probably men much inferior in strength and judgement, when all are properly fed are now able to do most work, because they have got the necessary quantity of food to support them. It may be asked, why did these men who are so much below the average get so unable to do as well as others? I would answer, through sickness, or being for some time unable to obtain employment, or inconsequence of the alteration of the ground being against them, or other causes.'

Sims' solution was to employ all men on adequate wages – usually 59 shillings a month, with boys paid proportionally less. This would, he believed, allow everyone to live reasonably well and to be kept fit for work. He would also employ competent overseers to make sure the wage was earned. He points out that this would remove the need for setting days, where the work was parcelled out. That he regarded as a whole day's work lost and the next day was usually taken off as well. This was all time lost as far as work was concerned. He goes on to mention other defects in the system, which were obvious a century ago; if the men are being paid for working in hard ground, there is no point in them working as hard as they can to clear more ground, because the price will be lowered. Conversely, if they are expecting soft ground and find themselves faced by hard, they will look for an easier way round it. So far he had only been discussing the tut system. He regarded tribute work as 'pregnant with evils'. His main complaint this time is against the men; 'It is a fact well known', that it was common for ' this class of miners to deceive the agents respecting the value of the ore ground, especially on the day previous to the setting, or on that day'. Once again, he is in favour of scrapping the whole system. Sims' words are especially interesting, because most commentators have sung the praises of tut and tribute, as encouraging competition and independent thought. His views do deserve consideration, if only because what he has to say was based on decades of practical experience. However, 15 years later, when the Commission reported the system was still in place, they saw little prospect of it changing in the immediate future. It is hard for anyone looking back on those times to judge the rights and wrongs of the argument but, in the end, how it is regarded depends on the point of view. The adventurers wanted to get as much work done for as little pay as possible, the miners the opposite. From the miners' point of view, however, the system did have one big advantage – they could dream of the time when they found that rich lode and for a short time at least made really good money. It was like winning the lottery, a lottery for which the wage owners could never have tickets.

It was not just how the money was earned that interested the Commission, but they also looked on the demands that were made on those earnings that limited what the family could and could not buy. Miners on subsist – and few were out of it for long – were in debt to the local shopkeepers and a few witnesses spoke of how they could exploit the situation. One said that overcharging was a big problem, quoting goods worth fifteen shillings being sold for twenty. Another pointed out that there was nothing the family could do; because of the debt held by the shopkeeper, they couldn't buy anywhere else. But more seemed to regard the shopkeepers as good men, without whose help they could scarcely have survived from day to day. The situation in regard to the company was rather different. They took money from the miners for a whole range of goods and services. Essential equipment such as candles had to be bought. At Cook's Kitchen in 1863, the total bill for candles for a group of six men came to £2 15s 6d, almost 5 per cent of the total value of the ore for which they were paid, and the company sold them at a profit. Because the amount the tributers were paid depended on the valuation of the ore, they also had to pay a share of the cost of bringing out and assaying the samples. Perhaps the weirdest item to be found in the list of expenses charged against the miners was the 'barber'. It seems some mines had a fixed charge of a few pence a month for a weekly shave by the mine barber, whether they wanted a shave or were growing a beard.

Another deduction that might be made was a fine, known as a 'spale'. There were various offences for which the men could be fined, such as not clearing away the dead material as they worked. One of the more common was for coming up early from their shift. This was a real problem for a conscientious miner, who almost certainly would not have had a pocket watch. His best system for timing was by counting the number of candles he had burned since starting at the face, but candles did not burn at a uniform rate, so it was easy to get that wrong. He would often have relied on his own body, fatigue telling him that the day was done, but that measure was even less reliable than counting the candles. The miners could get round the difficulty by waiting until they heard the next shift starting down the ladders, which only worked if the captain didn't spot them. The fine for coming up early was not necessarily matched by a bonus for staying down longer, though tributers who had hit a good lode might choose to do so themselves.

After looking at the working system, James Sims turned his attention to what he regarded as a vital topic: ventilation. As he pointed out, good ventilation was essential to the miners' health and, an argument to appeal to an employer, a healthy miner works better than a sick one. He was in favour of using the pumping engines to force air through the hottest parts of the mine. There were gradations in what might constitute 'bad air'. To some it was any air contaminated by dust or stagnating in dark, hot regions. To others it only qualified as bad when the candles were snuffed out for lack of oxygen. In the latter circumstances, the position of an individual deep underground could become perilous. A.K. Hamilton Jenkins in his book *The Cornish Miner* has a story about what happened to an old miner called Lisha Billing.

The men had all come up at ten o'clock one night, apart from Billing. Next morning, he had still not appeared and Mine Captain Boundey when he heard about it, suggested sending for his son, Andrew, who was an expert at locating lost miners. The young man set off with two other miners, and it soon became clear that Billing had lost his candles and had set off in the pitch black to try and find an exit. They found his footprints in the mud and followed him through a maze of levels, until they finally found him lying on the ground right in front of a deep shaft. His miner's instinct must have warned him of the danger and he had thrown himself down or he would have plunged to his death. It is salutary to think that in the mines a man's life depended on the flickering flame of a candle. Until the middle of the nineteenth century, the only way to light a candle was by using a flint and steel to strike a spark. This changed with the invention of the match – the 'lucifer' as it was originally known. This was a great boon to the mining community, but a disaster for the workforce, mainly young women, who produced the lucifers. The phosphorus at the top was highly toxic and resulted in a hideous condition known as 'phossy jaw'. The toxins attacked the lower jaw that usually had to be surgically removed. Candlelight was also improved by the introduction of the new tallow candles, produced by Palmer's of London. The tallow candles had, in effect, a double wick, wrapped round a metal core and as well as giving a brighter light than ordinary wax candles, they were less likely to be snuffed out. But even the best candles will only burn where there is a good supply of air and, in many parts of the mine, that was a major problem. The main difficulties occurred when the men were working far from the nearest shaft, where the air could be so bad that it was difficult to keep a candle burning. There were stories circulating of men having to take off their coats and wave them to create a draught of air before they could even get them to light. The men also often believed that a burning candle used up what little clean air there was to breathe, so they would snuff them out when they took their meal breaks, eating in total darkness.

There were ways in which the life of miners working underground could have been improved and made safer, particularly in the hazardous business of blasting. The system has already been described, and the inherent dangers of using iron rods for tamping. In 1814, Sir Rose Price had suggested adding a copper sheath round the iron, but the idea was seldom followed through. Indeed, one colliery manager on being questioned, agreed that it might save lives, but the rods were too expensive. Iron rods were still in general use in Cornwall in the latter part of the nineteenth century. The other problem was the unreliability of the primitive fuse system then in use. An answer to that became available.

William Bickford was born in Devon in 1774, but worked in Truro in the leather industry where he became aware of mine accidents. A friend, James Bray, owned a rope factory in Tolgarrick Road. Bickford realised that a strand of rope yarn impregnated with gunpowder could be included in a thin rope and would burn at a steady rate. He developed a machine for making his safety fuses, or 'safety rods' as he called them, and applied for a patent. He began manufacturing at Tuckingmill near Camborne. Sadly, Bickford died in 1834 just as the factory was about to open. In its first year of operation, it produced 45 miles of fuse. It was very easy to use, and the miner could simply cut off an appropriate length from a coil of fuse for each blasting. It was clearly a huge benefit, but the safety fuse was only adopted very slowly, partly it seems because of the expense and partly, no doubt, from a reluctance to accept change. But that safety improvement was badly needed. J.A. Paris, in a paper on accidents in mines through blasting published in 1817, described eighty accidents in just one area of Penwith, and although few were fatal, many men were blinded or seriously maimed.

The most arduous part of blasting was boring the holes, as described earlier, in which one man held the borer while another hit it with the hammer. But in the area round St. Just the work was even tougher as a mine captain from the St. Just area explained in his evidence to the 1864 Committee.

> 'And there is another thing, our men all work single-handed; we have not two men at an end here at a time. They hold the borer with one hand and beat with the other, and when the hand gets tired with the hammer the man changes it. That is why the eastern man cannot work with our men here. They must all go double-handed. Single-handed men get more money; two single-handed men in an end will get as much as three double-handed men. Single-handed men are the most paying to the adventurers.'

One can clearly see the advantage to the adventurers, who were paying the two men the pay of three, but getting the work of four.

Working between the surface workers and the miners were the shaftsmen. Their job was to ensure the pumps were always working correctly. As the pumps themselves were attached to the bottom of the pump rods, this involved going right to the bottom of the shaft to do the work. This was vital to the whole success of the mine. A typical large engine, such as the 80-inch installed at Consolidated Mines in 1832, raised over 33 gallons at each stroke from a depth of 265 fathoms below adit. The strain on the valves was immense, but they needed to be kept in perfect condition for the pumps to work. Tough leather was used, usually of buffalo hide, but at one mine the more exotic hippopotamus hide was preferred. These needed constant replacement, so at Consolidated Mines, the shaftsman would have to descend to adit level and then go a further 1,500 feet to reach a valve. Working in the dark confines of the shaft was inevitably difficult. But even if no repairs were called for, there was a constant need to make the journey down the shaft to check that all was in order, that all the holes at the end of the pump were below water level and not sucking in air, and to grease the moving parts. Most miners had their comrades for company – the shaftsman worked on his own.

The work down the mines was hard, as indeed was the work on the surface, and in the early years, conditions for making life more comfortable scarcely existed. Men might have been working deep underground in temperatures as high as 30°C and emerge to an icy wind of winter. In the early days, there were often no facilities provided for changing into dry, warm clothing. As mentioned earlier, the bal maidens also worked and rested in open sheds, but by the time the 1864 report on conditions came out, matters were greatly improved. The manager at Dolcoath described the facilities they had set up for surface workers:

> 'We have four mine establishments with an oven in each of them, large enough to contain two hundred pasties or hoggans. We have benches round a long room where they can sit down, and hot water is always prepared for dinner-time. As soon as the bell rings, they rush to get their dinner and so many as like to do so join together. Half a dozen perhaps purchase a tin kettle, and everyone brings a cup, and if they join together they get a little tea with the water.'

Things were improving in small ways, but in its essentials the life of the working miner in the early nineteenth century was much as it had been a century before. One other factor remained unchanged; pride in doing a job few who were not born to the life could manage. With it went a respect for physical strength. The miners also liked to pit their skills and strengths against each other, and although one would have thought they spent time enough underground drilling holes in hard rock, they also competed. A record was set when a group of three men, one holding the borer and the other two beating it down, bored a thirteen inch hole through granite in 6 minutes 43 seconds. It is small wonder that the men took pride in their achievements.

Chapter 12

SILVER AND LEAD

Because there are so many reminders of tin and copper mining in the landscape, it is easy to forget that mining for lead, silver and other minerals was once of great importance, but has left far fewer traces. Lead and silver were the most important and were found in different locations, the Tamar valley, Dartmoor and Combe Martin on the north coast of Devon. One of the most productive sources of the metals and among the earliest to be exploited were the lodes that lie in the Tamar valley, two running north to south, in parallel straight lines, south of Calstock, with a third set just above Bere Ferrers. The mines that worked these lodes were known collectively as the Bere mines. There were rich pickings, with the highest concentration of silver nearest the surface, in some places with a concentration of silver as high as 140 ounces per ton of ore. Because the lodes were relatively easy to access, mining began comparatively early, generally thought to be around 1290. Just as with other ores, it seems likely that men looking for good building stone dug down to the rock and discovered the interesting colouration and explored its properties. As soon as the miners began to have successes, Edward I claimed his customary rights of owning all the metal mines in the country and simply appropriated the silver. The miners could be paid for extracting and treating the ore, but there was no fortune to be made by selling it. And there was indeed a fortune to be made. In the first few years, silver flowed from the mines. An early record shows that between 12 August and 30 October 1294, 370 pounds weight of silver ore was sent to the king from Maristow Quay – today the quay is only just recognisable by the stone edging by the River Tavy. In 1297, silver valued at £4,046 – over £3 million at today's prices – was sent out, so this was a highly lucrative industry. But this was not mining on a large scale, as most of the valuable ore was close enough to the surface to be reached by shallow pits. There are signs of more elaborate measures being taken in the remains of a leat that stretched for some twenty miles and in places was carved by hand from the rock, having been built before the age of gunpowder. It was used to bring water to the mines.

As the king's coffers began to swell, it became clear that to increase production, mining techniques would need to be improved and that that would need men experienced in this sort of work. Lead mining had been carried out in Derbyshire since Roman times; the oldest known cast ingots seem to date back to the second century. One advantage of being a medieval king was that if you wanted something done, you simply ordered that it should be and it was. An edict was issued that called for 340 miners from the Peak District to set off for the long walk to Devon to work in the mines. Any who refused to go either had to find someone to take their place or be thrown into gaol. Whether the king felt diversity of ownership might be a good thing or not, he did agree to let one of the mines to the Frescobaldi company of Florence. It is not clear what they paid for the privilege, but they soon regretted the arrangement, as the ore was of poor quality, difficult to remove and expensive to transport. Perhaps the king or his advisers simply made a shrewd move – selling the Florentines a dud. The Florentines soon abandoned the region and once again everything was in royal control. The money continued to flow in and by the early part of the fourteenth century, silver worth £1,775 and lead valued at £850 was in the treasury. We do not know the full extent of the workings, but a reliable account of 1453 described the area of Bere Ferrers as having six mines 'reserved exclusively for the king's use.'

The labour force may initially have been pressed into service, but because of the wealth generated by the mines, the workers enjoyed higher than average wages. Soon there was no shortage of skilled workers, not only from Britain but also coming across from the continent. As well as bringing personal expertise, they also brought new ideas. It was said that they were responsible for improving drainage. The Ordnance Survey maps of today show a ridge between the Tamar and the Tavy rising to a height of around 400 to 500 feet. The new miners dug adits into the sides of the ridge, which enabled them later to dig shafts down to the levels above adit. Records show that gangs of men were paid a total of £12 10s for digging 'avidots' in 1297. As a result, men were able to work right through a wet winter as well as a dry summer and the production from the mines in the area doubled.

The output of the mines fluctuated wildly. A low point was reached in the mid-1420s when in a two and a half year period, the average yield of silver was only 12 ounces. But by the middle of the fifteenth century, output had shot up to an average of over 650 ounces a year. In the latter part of the century, the king's coffers were being swelled by around £40,000 a year, which could be worth as much as £100 million today. Not all of it came from the Bere peninsula. The Combe Martin mines were also prolific.

The earliest records show William de Wyndham beginning mining operations at Combe Martin in 1293. He was immediately successful and was able to send 370 lbs of fine silver to Edward I, who promptly handed it on to his daughter as a handsome dowry for her wedding. The records here also speak of miners coming down from Derbyshire, which suggests that they were indeed spread out over all the silver mines of Devon. Royalty continued to take the silver and use it for a variety of purposes – Henry V, for example, used it to finance his wars in France. It was Henry VIII who was the first for whom we have actual evidence of hiring German expertise, when he brought over Joachim Hochstetter, a German mining engineer, to take charge of the mines, with a work force of a thousand. The large number were needed because of the primitive conditions then in use. There was just one small water wheel to power pumps and the ore was brought to the surface via shammels – in other words via a series of stages, with the ore being thrown up by shovel at each successive stage. German technology at this time was far more advanced. The one reliable account of mining practices in Europe at this time is Agricola's *De re Metallica* published in 1556, and it contains very full accounts of the mining practices as he saw them in Joachimsthal in what is now the Brandenburg region of Germany. This was an area famous for its silver mines. It is clear from the descriptions and many illustrations that everything was very efficiently organised. Various types of water-powered pumps are described, and the horse gin for lifting material up and down shafts was illustrated. No doubt, Hochstatter brought many new ideas with him when he came to Devon.

The lease on what became known as the Fayes mine was acquired in 1587 by Adrian Gilbert, and the following year he went into partnership with mines expert Bevis Bulmer. The latter was to pay the working costs and supervise the mine and they were to take an equal share of the profits. It turned out to be an excellent deal, with the partners making £10,000 each in the early years. It was an immense undertaking and records show a shaft 32 fathoms deep, and 32 fathoms wide. In a report written in 1835, when a group of adventurers were working to restart operations by working down below the old levels, they estimated that a ton of ore from the original mine would contain anything from 54 to 100 ounces of silver and 14½ cwt of lead. Work, however, was still limited by water, and the miners were never able to get below 50 fathoms, without being inundated. As in other mining regions, progress now depended on the arrival of new technology. Even that did not help at first. As the low level workings became less profitable, attempts were made to restart the mines in 1780 and 1813 but, in both cases, they were abandoned, as the waters swept back in.

Whichever area they were working in, these miners earned more than other workers in other industries, and because they worked for the royal family enjoyed special privileges. They were excused from a variety of taxes and in general were not answerable to the ordinary law officers, but only to a warden. Punishment for offences were less draconian than for many of the poor, largely, no doubt, because it was in the wardens' interest to keep the men at work rather than languishing in a cell.

These early successes in Devon did not last and work on the Bere peninsula also seems to have dwindled to the point where there are scarcely any records. One thing we do know from this period is that ore was smelted locally, but quite crudely. Where other metals were smelted in furnaces in which the temperature was raised using water-powered bellows, here simple kilns were built into hillsides. Positions were chosen where they could catch the prevailing westerly winds to increase the heat of the fire. Change came slowly; gunpowder was first introduced in the middle of the seventeenth century.

Revival of the mines in the Bere region stuttered along for a while. The primitive smelting methods meant that a lot of valuable silver was lost and thrown out with the slag and there were attempts to recover it in Elizabethan times, but an effort to reopen workings at the end of the seventeenth century came to nothing. A century later, Christopher Gulett of Tavistock took over working the South Hooe mine on the Tamar west of Bere Alston. He was deputy clerk of the peace for Devon and an inventor, who patented a hydraulic engine in 1773. A Christopher Gullett also took out a patent for a miraculous cure for gout that required no medicine nor any form of plaster. Sadly, the mining Gullett was not quite that versatile; the gout remedy was the work of a Doctor Gullett from Exeter, but presumably of the same family. We do know, however, that the mine proved exceptionally rich with 140 ounces of silver per ton of ore and in 1784 and 1785, the mine produced 6,500 ounces of silver. The rich lode ran right under the river, which would have made it inaccessible to earlier miners, but by this date the powerful and efficient Boulton and Watt engines were available for pumping. The area was about to enter a new age of prosperity.

The most important developments in the Bere region in the early nineteenth century were largely down to Percival Norton Johnson. He was unlike most who managed mines in that he came from a very different background, as a London based assayer. He was born in 1792 and was apprenticed to his father at the Worshipful Company of Goldsmiths. By 1816, he was a liveryman of the Company with his own business in the jewellery district of Hatton Garden. He made a number of notable advances in metallurgy, including developing an alloy of nickel, copper and zinc that was known as British plate and widely used for manufacturing cutlery that looked like silver. He wrote a number of scientific papers and was made a Fellow of the Royal Society. He later went into partnership with a stockbroker 'John Matthey' and the new company developed a technique for refining platinum on an industrial scale. The company continued in existence after Johnson's death and is still a leading innovator as Johnson Matthey. He could well have spent the rest of his days in London with his assaying business, but his travels took him to Devon and Cornwall, where he decided to invest in mining.

Johnson decided to develop the rich ores that lay along the western lode and stretched out under the Tamar riverbed. His idea was revolutionary. Instead of sinking a vertical shaft, he created a 25° incline. Men no longer had to climb ladders but could walk up and down the comparatively gentle slope. The incline was continued out under the river, until it had reached a depth of 125 fathoms. This was too much for the haulage engine at the surface to cope with, as the incline itself was almost 300 fathoms long. So, a cavity had to be excavated from the solid rock and a second engine of 20 horse power installed at a depth of 115 fathoms below the river bed. A vertical 60 fathom shaft was dug from the engine to the lowest level of the mine. The exhaust fumes were carried up a flue to the surface. The whole system worked well and the incline ensured far better than average ventilation; too good on some days, when the powerful draught would blow out the miners' candles.

He also introduced a device that had been developed in Germany for separating out the ore. This was the shaking table. It consists of the table itself, covered in fine corrugations, and held at a slight slope. The ore is introduced at the top as a slurry together with wash water, and as the table is vibrated the heavier minerals settle out at the top, the lighter are washed down towards the bottom of the table. Two tables of this type can now be seen at Geevor Mine. The result of good management and innovation was that the mine prospered. The ore deposits were rich, usually 50-60 ounces of silver a ton, sometimes even higher and as a result, dividends were high. By 1852, the mine was employing around 200 workers and the original investors who had put up £12,000 received dividends of £44,000 in 1852.

Unlike many stories of fortunes being made from mines and benefitting a few, this one is refreshingly different. Johnson was a philanthropist. He built decent cottages for the miners' families and schools for the children. He was also aware of the many miners who had been blinded in accidents and decided with his wife to do something about it. She was one of the prime movers in establishing a charity – the London Society for Teaching the Blind to Read, while her husband developed a method of printing using embossed letters that could be read by touch – a precursor of Braille.

As a metallurgist, Johnson inevitably took an interest in the smelting process. His company acquired works at Weir Quay on the Tamar in 1845. By 1850, the works had been modernised with eighteen furnaces capable of smelting 300 tons of ore a month. The ore was first calcined and then subjected to a higher temperature to separate out the silver. The facilities at the quay were improved to take vessels of up to 400 tons that enabled the smelters to accept ore from as far away as Newfoundland. Although it is hard to imagine the scene today, there is still an active boatyard on the quay. Johnson was always on the lookout for the latest improvements and, in 1850, the works started using the Pattinson process. Hugh Leigh Pattinson had patented his idea in 1833, which was based on the fact that molten lead with the lowest silver content would be the first part to solidify out on cooling. A typical system consisted of nine iron pots. The central pot was filled with ore and heated to melting point. On cooling, the top portion was skimmed off and added to the next pot at the side, while the remainder was added to the pot on the opposite side. The process was repeated and at the end of the process, there would be lead at one end and silver at the other. It was claimed that 98 per cent of the silver could be removed in this way. In 1852, the Tamar smelting works were sold against Johnson's advice; it appears that short term gains were preferred to long term income. Johnson retained a smaller smelter which showed the advantage of taking a longer view by remaining in business right up to 1898, by which time silver mining in the region had more or less ended. When Johnson died in 1866, the loss of his guiding hand was felt and the works entered a period of slow decline. He was buried in Stoke Fleming, where he had made his home, and a memorial window was installed in the local church. He deserves to be remembered as a successful metallurgist, an innovative mine manager and, above all, as a humane benefactor to the local community.

The only other mine to match South Hooe was South Tamar Consols, just south of Weir Quay. The mine was set above the eastern lode, and had last been worked in the reign of Elizabeth I. A company was formed in the 1840s to open up and extend the old workings, and, by 1846, Cowie's shaft had been sunk to a depth of 90 fathoms and workings had spread over half a mile under the river. The investors were rewarded with handsome dividends, sharing out £32,000 in the first ten years. Like South Hooe, this was an extensive concern, with eight steam engines, employing 150 men and with over ten miles of workings close to and under the river. But everything came to an abrupt end on the evening of 31 August 1856. A fault had been plugged with clay and gave way. The river rushed in with such rapidity that the air was forced out of the mine, blowing the tops off the covered shafts. The only good news was that it happened on a Sunday when no one was at work. At any other time, any men working underground would have inevitably perished. Various plans were put forward for reopening the inundated mine and one report claimed that Isambard Kingdom Brunel was even called over from his mighty project on the Tamar bridge at Saltash to give his opinion, though he could hardly be considered an expert on mining. In the event, nothing could be done and the mine was abandoned. It was equally bad news for the adjoining East Tamar Consols, which managed to continue in production by working the Lockridge mine a little further north. But by 1862, the pumping engine could no longer cope with the rising river water and the mine was abandoned. By the end of the nineteenth century, it was all over and seven centuries of mining for lead and silver in the Bere peninsula came to an end.

Engine houses proliferate in Cornwall and to a lesser extent in Devon, but only one survives on Dartmoor – Wheal Betsy, just north of Mary Tavy. There was a mine recorded here in the 1740s, but that closed down. It was reopened in 1806, together with nearby Wheal Friendship. It closed again in 1846 and then had a final lease of life as Prince Arthur Consols, the name it retained from 1863 to final closure in 1877. The engine house stands by the Job Shaft, which reached down to the lode that runs from north to south through the site. In its heyday, there were four other shafts. The surviving engine house was home to a pumping engine. This is a remote and exposed site and water for the mine had to be brought a considerable distance. At the beginning of the nineteenth century, the Reddiford leat was constructed to bring water from the Tavy. It followed the contours of the land, dropping just 26 feet in its 4½ mile length. Today it ends at an odd linear reservoir, named after what was once the adjoining tin mine of Wheal Jewell. Originally it continued all the way to Wheal Betsy. The mine had a somewhat chequered history, but in its time it provided a considerable quantity of silver: 2,056 ounces between

1855 and 1859; 4,727 ounces, 1865-7; and 2,019 1870-76. The engine house is now preserved by the National Trust.

Manganese was produced in considerable quantities in Devon. Its main use at first was in purifying glass and in ceramics, though in the latter part of the nineteenth century it was to become a vital ingredient in steel manufacture, following the introduction of the Bessemer furnace. One of the first accounts of the industry comes in Thomas Moore, *History of Devonshire*, 1829. The earliest records he discovered relate to the opening of a mine at Upton Pyne, just north of Exeter in 1770, closely followed by two others at nearby Newton St. Cyre. They were productive for a while, and in the first decade of the nineteenth century, 2-3,000 tons a year were being shipped out from Exeter. By 1810, however, they were largely exhausted, and production shifted west of Exeter to the area round Diddescombsleigh which also did well for a while but by 1821, production had fallen to 450 tons.

As the mines in the east of Devon began to fail, new deposits were found in the area round Tavistock and these proved the richest of them all. The mines lay close to the Tavistock Canal and ore was brought down to the quay at Morwellham – 1,335 tons in 1819, 2,212 tons in 1821. Morwellham Quay is one of the few places where one can find any reminders of this once vital industry. The ore was crushed in a mill, using two rollers powered by a 32-foot diameter overshot wheel. This was later removed but was recovered and restored and is now a striking feature of the quay. The powdered ore was packed into barrels, made at the cooper's shop on site and despatched from the special Manganese Quay. Manganese remained an important trade, but it was later to be dwarfed by the output of the Devon Great Consols mine in 1844, and once that was in full production it was said that there was never less than 4,000 tons of copper ore on the quay at any one time.

There were other minerals worked in small quantities, including iron pyrites, but they were never significant. Silver, lead and manganese may never have been produced in the quantities of copper and tin, but they were nevertheless important contributors to the economy, especially of Devon. Two other elements were extracted on quite a large scale, arsenic and tungsten, but we shall be looking at that in later chapters.

Chapter 13

THE COUSIN JACKS

'Cousin Jack' was the name given to Cornish miners working overseas and the name appears all over the world from Australia to South Africa to Latin America. As one writer put it in the nineteenth century, 'Wherever a hole is sunk in the ground today – no matter in what corner of the globe – you will be sure to find a Cornishman at the bottom of it, searching for metal.' And they were all Cousin Jacks, though no one seems to know the origins of the name. There is no shortage of suggestions: Cornish miners wanted a job for their cousins; if they couldn't do a job themselves then Cousin Jack could and so on. But whatever its origins the name stuck.

Anyone with an interest in the miners of the South West will know of the great Cornish diaspora but my own first experience of finding the physical evidence was quite close to home. For years, the family had regularly holidayed in Cornwall and my wife and children had got used to the fact that the sight of a previously unvisited engine house would mean an inevitable delay, while I wandered around, noting any interesting features that distinguished it from other similar buildings. It might be fascinating for me, but simply meant a delay in getting to the beach for everyone else. So when we decided on a change of scene and headed off to Bantry Bay in the South West corner of Ireland, there was a general sigh of relief that there would be nothing to distract father. As we got near the little fishing village of Allihies, there it was on the top of the hill – the unmistakable outline of a Cornish engine house. We had inadvertently found ourselves close by what had once been one of the most productive copper mines in Ireland. Of course, the building had to be investigated. The engine house was absolutely typical of Cornish practice, but the surprise was to find a gaping shaft still exposed right next to it. The site itself is rugged and rocky, and transport to and from the mine must have been extremely difficult.

The Allihies mine was worked from 1812 to 1875, one of three important producers. The others were Knockmanon in Waterford and Avaca in Wicklow. Records show that the Allihies mine was first set up by twelve Cornish miners under the direction of mine captain John Reed, while work at Knockmanon was carried out by another Cornishman 'John Petherick' and his team. Allihies was a great success. Between 1835 and 1844, 52,586 tons of ore were exported and the mine was employing 1,200 workers. The port of Castletownbere was developed specifically to deal with the output of the mine. A large percentage of the workforce was employed in carting the ore.

There is no need to describe the way in which the mine was worked as it was essentially the same as it would have been in Cornwall. Conditions certainly were very similar. An intrepid, but anonymous, writer who was touring round Ireland by yacht, visited the mine, and from his account probably regretted it. His story was reproduced in an article by Des Cowman, *Life and Labour in Three Irish Communities c. 1840,* 1983. It is worth quoting, if only to point up how similar his experience was to those of visitors to Cornwall. The descent of the 120 fathom shaft was by vertical ladders: 'some of the rungs were wood, some iron, some broken, some entire and all more or less slippy'. Worse was to come:

> 'With backs bent to very nearly a right angle, with devious footsteps at no time on dry land in the full sense of the term, but often indeed up to our knees in pools, with a splash which spread up acherontic mud into our eyes and mouths placed conveniently to receive it, with a continual drip from the oozy roof.'

And so on it went on with increasing misery, on what was intended to be a pleasantly interesting interlude in his yachting progress.

It is not clear how many Cornish were employed and how many were locals at any of the mining sites, but it is evident that there was a sudden explosion in the population of all three mining areas. At Bunhamon, the nearest town to the Knockmahon mine, a report of 1832 described wretched housing conditions, where houses built originally for a family now held three to four families in every room. In 1836, the mine owners built sixteen houses and more were added later. They would have liked to build even more, but the local landowner would not sell the land, because, according to the parish priest, he did not want 'the labouring class' on his land. The priest deplored the lack of proper accommodation, not on the obvious grounds, but because he felt it might lead to immoral behaviour. Elsewhere, conditions were no better. At the Avoca mine, 410 people were recorded as living in 63 houses on a sloping site right next to the mine.

In the early years, there were numerous complaints about drunkenness. The customs authorities at Castletownbere reported 14,953 gallons of whisky were drunk in a single year – and the figure obviously would not include any distilled illegally. A temperance movement developed in latter years and was surprisingly successful. The mines themselves were subjected to fluctuating use and a sharp fall in the price of copper in 1845 led to temporary closure, with many workers laid off. The biggest drop in prices came with the opening up of exceptionally rich copper mines at Butte, Montana – and many of the workers at the Irish mines packed their bags and left for richer pickings. This was a period, too, when many Cornish miners were heading abroad, but the movement across the world began at a much earlier date. Around the world, there were mining concerns that needed not just Cornish expertise but also Cornish technology.

In 1813, a Peruvian called Francisco Uvillé called on Richard Trevithick. He had arrived in Britain in order to purchase machinery for a company in which he was a partner. It had been formed to work the silver mines of Cerro de Pasco set on a windswept plateau at 14,200 feet above sea level, between the eastern and western ranges of the Cordilleras. Silver had been found there in 1630 and the town received the popular name of 'El Opulent', but there was very little that was opulent or organised about the town or the mining. Lacking technology, the miners dug a pit as far as they could go, then abandoned it and started a new hole somewhere else. The area was as full of holes as a giant gruyere cheese, among which were scattered mean little houses, with thatched roofs and windows blocked only with paper. The opulence of the mines seems not to have been used to benefit the lives of the miners. Uvillé had come to Britain to buy steam engines and pumps to enable the mines to be developed profitably and efficiently. He had begun in the obvious way by consulting Boulton and Watt, but must have been alarmed at the size of the powerful engines and would have puzzled how he could ever get them up the mountains. Then he received the dispiriting news that at the altitude of the mines, the rarefied atmosphere meant that they would probably not work anyway. He returned to London where he saw the model of a steam engine in a shop near Fitzroy Square. He asked for details and was told it was a model of a high pressure engine designed by Trevithick. It was at once obvious that this could be the answer he had been looking for; the high pressure meant that the engine was smaller than a Boulton and Watt of equivalent power and it would work at altitude. He took a ship to Falmouth and went looking for the Cornish engineer.

Uvillé had been authorised to spend up to $30,000 on acquiring two steam engines with pumps and to hire two workmen to return with him to set them up. He was specifically ordered not to offer any form of partnership to anyone in Britain. But when it came to negotiations, he could hardly be expected to refer decisions back to Peru and wait for a reply that could take months. So, encouraged by Trevithick he ordered not two but six engines. Trevithick ordered work to begin on preparing the six engines, each with 24 inch cylinders, and each to be supplied with 25 fathoms of 12 inch pumps and 10 fathoms of 3 inch water supply pipes. They also required the necessary high-pressure boilers. Everything had to be made in sections for transporting and then reassembled on site. There was a problem with getting the money through to pay the workmen who were preparing all the machinery, but Trevithick put his trust in Uvillé and paid £3,000 for a $12,000 share in the mines that would entitle him to one fifth of the profits of what was potentially the richest silver mine in the world. When everything was finally ready, two men were chosen to go with Uvillé to oversee the work, Thomas Trevarthan with William Bull as his assistant. It was felt that a third man would be helpful and the job went to Trevithick's brother-in-law Henry Vivian. He was a qualified engineer, but Trevithick was concerned that he was prone to 'making too free with an evening glass'. The party set out for Peru on 1 September 1814.

The Cornishmen must have been horrified when they saw the route to the mine up which they would have to go with all the machinery carried by a string of mules. A traveller along this route, Hugh Salvin, described the journey as he made it in the 1820s.

> 'The perils of the journey were very considerable. He had to pass over deep ravines, along bridges not that more than five or six feet wide. The passage along the precipitous sides of the mountains were still more frightful. The mule always walks close to the edge of the precipice, and at times his leg was dangling over a descent alarming to look at.'

It was up tracks such as these that Uvillé and the three Cornishmen had to make their way with the mules. Everything had to be in sections of a size and weight that could be managed by the animals, so they had to weigh no more than 300 pounds and not be more than seven feet long. Once they had arrived, they had to be assembled under the direction of Uvillé who had been given a crash course in engineering by Trevithick, two ordinary Cornish miners and an engineer who was liable to get drunk at any time. They had the help of a workforce who had never seen a steam engine in their lives and with whom three of the four supervisors were totally unable to converse. It was not an arrangement that was ever likely to succeed. Nevertheless, one engine was erected – an event celebrated in a Report of the *Lima Gazette* of 10 April 1816:

> 'Immense and incessant labour, and boundless expense, have conquered difficulties hitherto esteemed altogether improbable; and we have with unlimited admiration, witnessed the erection, and astonishing operation of the first steam engine ... we anticipate a torrent of silver, that shall fill surrounding nations with astonishment.'

But Trevithick was getting worrying reports from Peru of things going badly wrong and decided that given the difficulties of communication with a country so far away, the only sensible thing was for him to make the journey to Peru himself to sort things out. He reached Lima in February 1817 and received a very flattering welcome from the local paper, who described him as Don Ricardo and awarded him the title of Professor. As well as providing engines for the mines, he had also supplied one for the mint in Lima, which was now working well, and when he finally reached Cerro de Pasco he found there were still two pumping engines waiting to be installed and that the ones that were working needed to be put to rights. He set about the task with great optimism since, as he wrote in an undated letter to England, one mine was 'immensely rich' and others would soon be opened up producing 'a much greater quantity'. Already, he was accumulating a considerable quantity of bullion in his name as his share in the partnership. Things seemed to be set for a prosperous future. However, Trevithick may have known a great deal about mines and steam, but took little interest in politics. Had he done so, he might have been aware that there was growing revolt in South America against Spanish colonial rule and that the rebellion was about to overwhelm Peru.

A liberation army led by Simon Bolivar moved into Peru and the Cerro de Pasco mines were recognised as a valuable prize, though not by everyone. First, the Royalists moved in and began breaking up the engines to provide iron for horseshoes and spurs. Trevithick managed to have many parts buried away for later use but when Bolivar himself arrived, everything changed. He recognised the importance of the mines as a valuable asset and appropriated everything, including all the ore already raised. That of course included Trevithick's own share that was awaiting shipment to England, estimated to be worth £24,000. He would have made a fortune, but all was lost. Even worse, Bolivar 'recruited' him into his revolutionary army to design guns for them. The Cornishman was to spend eleven years trying to make up for the money he had lost, but never succeeded. He was not alone in finding that venturing from Cornwall to South America could be a hazardous affair.

Gold had been mined commercially in Mexico under Spanish rule, but the methods used were outdated and most of the mines were in such a poor condition as to be more or less worthless. However, in the 1820s, a Cornish consortium bought up mines and became the Gentlemen Adventurers of the Mines of Rio Del Monte. Just as Trevithick had done, their objective was to bring steam technology from Cornwall to South America. In 1825, a party of Cornish miners set sail from Falmouth with 1,500 tons of mining equipment. When they arrived in Mexico, they were turned away from the port and were forced to land all their material on a beach near Veracruz. There were no roads from there to the mine and as the party began pushing inland through a morass of swamp and jungle with their huge quantity of equipment, the rainy season arrived, bringing with it mosquitos and the deadly yellow fever. Thirty of the Cornish died, together with a hundred of the Mexican labourers who had been employed to help. The rest of the party retreated to the hills away from the fetid swamps and had to wait for months before continuing their journey. Now they had to make their way to the mine site, which was not only 250 miles away but stood at an altitude of 8,900 feet. They had to make their own roads as they went and the whole journey took 14 months. John Buchanan, an engineer who made the great trek, recorded in his diary, 'After great labour and many accidents we conquered the great ascent and our convoy proceeded on our last stage to deposit its valuable cargo'.

The effort proved worthwhile. The mines proved rich in both silver and gold. In the 1830s and 40s, some 350 Cornish arrived at the mines. Unlike Richard Trevithick, one at least made a fortune. In 1858, Francis Rule arrived from Cornwall and did so well that he became known locally as 'El Roy de la Plato' – King of the Silver. Among the buildings he personally paid to have erected was a Methodist church. Other reminders of those days are four surviving engine houses – and a game. It was the Cornish miners of this district that introduced the Mexicans to football.

Robert Stephenson, who was to become one of the great railway pioneers, was also destined to spend some time in South America, grappling with the difficulties of the terrain. He had been working with his father on surveying the Stockton & Darlington Railway and then on the plans for the

Liverpool & Manchester, when he received an offer from the Colombian Mining Association to take over the management of a silver mine high in the Andes. There has been some speculation as to why he took the job. L.T. Rolt in his biography of the two Stephensons suggests he might have fallen out with his father, but it seems equally plausible that to a young man – he was just twenty-one – it would have appealed as a chance to be independent and have an exciting adventure in a strange land.

He left Liverpool in June 1824 and eventually arrived at La Guiará in what is now Venezuela but was then part of Gran Colombia. His assistant went on to Cartagena with the rest of the party and the mining equipment. Robert's job now was to see if it was going to be possible to use the port for shipment, preferably by connecting it to Caracas by a railway. The start was not promising. The ship had to anchor in the bay and all cargo had to be unloaded into lighters. Robert suggested that this could be improved by building a pier that ships could lie alongside for a modest £6,000. His first impressions were not favourable as he 'observed in silence the miserable appearance of the town'. As for building a railway, one look at the steep, forested hillside convinced him that it was wholly impractical, and even if it could be engineered the cost would be prohibitive. In fact, a railway was built but not until 1883 when technology was much more advanced. Robert's next assignment was to head for Bogota and to look out for any potential mine developments along the way. He set off with a servant and an interpreter to make his way along the mule tracks to climb the hills and head for Bogota, a journey that took him four months. Meanwhile, his assistant Charles Empson and a group of Cornish miners had made their way up the Magdalena River to the head of navigation at Honda. And there they were stuck with all the heavy equipment. It was the same problem that had plagued other expeditions – although the first part of the route to Maraquita lay across a level plain, the final twelve miles to the mines at Santa Ana were along a narrow mule track climbing the steep slope of the Andes. Robert sent word back to England that there were transport problems, but the next assignment of machinery arrived before they could respond.

Robert made his way up to the mines, where the mountain air was a great relief after the stifling heat of the plain, and had a house built for himself. While Empson set about setting the mine to work, Robert went off on a thousand mile journey by mule looking for potential mine sites, a lonely existence, and he wrote to his step-mother that he often thought of 'an English fireside and all the joys that spring around it'. At the end of his journey, he had to return to Maraquita to meet the main party of Cornish miners. It did not start well. The men had been recruited with the promise of high wages, but found themselves in a place where the only thing to spend their money on was booze, and they proceeded to do just that to the engineer's despair.

> 'I have no idea of letting linger out another week without some work being done. Indeed, some of them are anxious to get on with something. Many of them, however, are ungovernable. I dread the management of them. They have already commenced to drink in the most outrageous manner. Their behaviour in Honda had, I am afraid, incurred for ever the displeasure of the Governor, at all events as far as induces me to calculate upon his friendly co-operation in any of our future proceedings. I hope when they are once quietly settled at Santa Ana and the works regularly advancing, that some improvement may take place.
>
> 'It appears remarkable that having been all my life accustomed to deal with miners, and having had a body of them under my control, and I may say in my employ, that I should now find it difficult to contribute to their comfort and welfare.'

His lack of control puzzled Robert, but is not altogether surprising. The Cornish respected their mine captains, who were experienced at mineral mining and were familiar with their way of life. Here they were confronted by a young man whose background was in coal mining – an area of expertise that did not impress them very much. And although Robert had hoped that when they reached the mine site things would improve, with temptations removed, new temptations soon appeared as local traders saw an opportunity to cash in on the newcomers and continued to make sure they would never be short of a drink. Drunkenness remained a major problem. And he never did get a full day's work from the whole group at the same time. He estimated that most men averaged at best half a day's work each in any one week. And when they were drunk, they were always likely to pick a fight with someone, if not Robert himself then with his assistant, Empson. Relations did improve with time, but the mines never proved as profitable as the Adventurers had hoped. Robert was now desperate to get home. He handed over responsibility of the mines to the agent in Bogota and headed for Cartagena and one of the most remarkable coincidences in the history of technology.

When Robert arrived at the port, he was told there was another English engineer there, who had suffered great losses. This was Richard Trevithick, who had at the end of his eleven years in South America opened a gold mine, but in his attempts to find a route from the mine to the sea had contrived in a terrible journey to lose all his possessions and was almost consumed by an alligator. The two men met and there could scarcely have been a greater contrast in their prospects. Trevithick was going home without the fortune he had hoped for; Stephenson was returning to start a brilliant career, which would include designing one of the most famous locomotives ever built, *Rocket*. Meanwhile, the Cornish stayed on in South America, many settling permanently. Many others signed up for shorter terms and far from being the drunken louts portrayed by Stephenson were sober workers who took care that their families were provided for back in Cornwall. In 1834, John Chynoweth of St. Agnes went to work as a stampsman for six years for the Colombian Mining Association at a monthly wage of nine pounds. He appointed two trustees in St. Agnes to receive half his pay and to give two pounds a month to his wife for her and the children, but they were authorised to spend extra on medical care if anyone became ill. The remainder of the money was lodged with a savings bank for when he returned. Many others made similar arrangements.

These early adventurers abroad went by choice, lured by the prospects of good pay, generally at rates far higher than those in Cornwall. In the second half of the nineteenth century, however, they were driven by necessity. Copper had been the main earner for some time. In a report on Cornish mines published in 1838, it was estimated that the value of tin coming out of Cornwall that year was £462,000 while copper was worth over twice as much. But the copper ore production was reaching a peak, though it could have continued as a very profitable enterprise if the rest of the world had not suddenly started producing the metal in vast quantities. Some came from South America but the most important producer was the state of Michigan in America, where quantities of good ore had been found in an area to the south of Lake Superior. The result was a catastrophic fall in the price of copper and many of the old mining areas saw widespread unemployment. Families had to rely on public relief and from 1865 to 1867, public relief funding rose by 50 per cent in the St. Austell and Redruth area, but by 100 per cent in the St. Just area. For the Cornish miners, the answer seemed obvious; if they could not mine copper in Cornwall, then they would mine it in America. In that same period, some 600 men left to work overseas, recruited from just the three districts of St. Austell, Helston and St. Just. The early story of copper mining in Michigan is also the story of Cornish workers – the Cousin Jacks.

A.L. Rowse, in his book *The Cousin Jacks* (1969), described the arrival of the Cornish in the region and gave accounts of some of the men and what they achieved. John Hunt, who was born in Camborne in 1817 and was one of those who had gone to Ireland, arrived at Copper Harbor on the shore of Lake Superior in July 1846, having been contracted to open up a mine on the Keweenaw peninsula that juts out into the lake. Until the mining industry arrived, this was a thinly populated area – and when Hunt arrived at Copper Harbor he found there were just three families living there, so he had to return to the boat to get somewhere to sleep. He moved on and opened up the Boston mine, but as it seemed unlikely to prosper, he went down to Pennsylvania to the coal mines – an unusual step for a Cornish mine captain. He was soon back in Michigan, where he was in charge of the pumping and mining operations at North West Mine and after that was appointed mine Captain at the Cape Mining Company. In 1859, he set up a business of his own at Houghton and won a contract to build a tramway from the Quincy Mine to the stamping mill on Lake Portage. It was a double tracked line about half a mile long, but enabled the broken up copper ore to be brought straight from the mine in trams and tipped directly under the stamps. He was later to join his brother, Richard, who had gone north to Canada in 1854 and they bought a steamship, *Ivanhoe*, and started the Overland Transportation Company, which was mainly engaged in hauling copper from the mines to the lake for transportation. They had over 100 teams engaged in transporting copper ore. They made a major contribution both to mining and mine transportation.

William Harris from Illogan in Cornwall was another of the pioneers who helped develop copper mining in the Lake Superior area. He began by opening up the Bruce Mines in 1846 on the Canadian side of the lake. Miners were recruited from Cornwall and the company at once set about building docks and warehouses. Two of the managers had houses at the entrance to the docks to control traffic. Only boats approved by the company could land. This ensured that all the goods that came in were sold through the company store at whatever price they chose to set. In 1850, Harris moved south to become mine captain for the Minnesota Mining Company. It was during his time there that an immense mass of native copper was discovered weighing 500 tons that took forty men twenty months to cut up into sizes that could be transported from the mine. Harris went on to become agent for the company, before setting up in business. He was a prominent citizen and served in the Michigan legislature from 1871-75. The miner from Illogan ended his days as the Honourable William Harris.

There are other stories of Cornish pioneers who prospered, but for the ordinary miners who came over, life was similar to that at home in some ways. They still worked to the old systems they had been used to, arguing and bidding for work. But this was not Cornwall and in some ways it was more like the Wild West of popular mythology. The trouble began when Irish workers appeared in the wake of the Cornish. In the absence of any recognised law authorities, things could move from quarrel to tragedy quite swiftly. Henry Hocking had established a profitable bar at Front Run on the Ontonagon River, when an Irishman 'Patrick Dolan' set up a bar nearby on what Hocking regarded as his land. When Dolan was absent, Hocking set out with an axe and demolished the rough and ready saloon. When Dolan returned and started to rebuild, Hocking confronted him and demanded that he got off his land. Dolan, not unreasonably – if unwisely – told him that if he had a legal right, he should enforce it with the law. At which point, Hocking settled the matter by pulling out a pistol and shooting the Irishman dead.

Fights between Irish and Cornish workers were equally likely to end in tragedy. When an Irishman killed a Cousin Jack with an axe, the Cornish retaliated by burning down an Irish store. Although such lurid stories were not altogether common, they did not represent the norm for life in the mining district. Most were working too hard to have much time for anything else, and sudden death was still more likely to come from an accident underground than from murderous attacks.

The Lake Superior mines prospered. Statistics from the *Journal of the American Geological and Statistical Society* in 1859 show that when work got under way in 1845, just 11 tons of copper were produced, but by 1858 this had grown to an impressive 46,307 tons. As the copper was priced at $500 a ton, this brought in well over $2 million in revenue – an excellent return on investment. Copper was profitable but the profits went to the investors and although the Cornish miners may have had better wages than they would have received at home, they were not going to make their fortunes. But that possibility did exist further south.

In January 1848, John Marshall, who was a partner in the water-powered Sutter saw mill on the American River in California, was inspecting work that had been done the previous night to widen the channel for the tail race. He spotted shiny specks in the silt and as in his own words he had some knowledge of minerals, he knew they could be one of two sorts – iron pyrites, fool's gold or the real thing. Both look alike, but iron is brittle, gold malleable. Tests soon showed that he had indeed found gold and as word spread around the world, the great California gold rush began. Many Cornish were among the famous forty-niners.

Hundreds of miners left Cornwall for the difficult journey to California. There were many ways of getting there. In just one week in April 1849, 1,000 left for Quebec to make their way down to California. The fastest route, however, was to get to New York, take a steamer down to Panama and then make their way across the isthmus and up the west coast. Many miners already at work in the copper mines in the north decided to try their luck in the gold fields. Edward Dale was a miner from St. Agnes who sent a description of his journey with a party of fellow Cornish workers. Things went well at first but when they reached the Platte River in Nebraska, they were told that there were so many trying to make the same journey that all the ferries were already fully booked. Being resourceful men, they simply built a boat of their own and not only got themselves across but made extra money by ferrying others. It was a bold enterprise as the river is fast flowing and about a quarter of a mile wide. They later crossed the Humboldt Sink, where Dale's ox that had hauled the wagon died. But Dale himself made it through and joined the tens of thousands of others who were busy panning for gold.

In some ways, it was reminiscent of the early days of tin mining, where streaming was followed by the sinking of deep mines down to the veins of metal. It was here that Cornish expertise, knowhow and equipment was most valuable. The most prolific mining area was Grass Valley in Nevada County – not to be confused with Nevada State. Here, once again, there was a certain amount of conflict between the experienced miners from Cornwall and the less experienced but eager Irish workers. One of those who came to Grass County was a Cornishman, Philip H. Paynter, who had previously worked at Sutter's Creek. Like the other Cornish, he used the technology that he had known back in England, including blasting with black powder. Then, in 1867, the Giant Powder Company was formed in San Francisco to manufacture the new explosive patented by Alfred Nobel – dynamite. The Cornish did not want to use it, preferring to trust the well-tried methods of old. They had some cause, as the first Black Powder factory was actually destroyed in an explosion in which workmen were killed. Paynter organised a strike and the absentee owners in San Francisco sent in Chinese workers to try and break it. At this point, the Irish formed a united front with the Cornish and the owners backed down. The result was the formation of the Miners' League, a trade union with Paynter at its head. Grass Valley became known as Little

Cornwall, and the road nearest the spot where gold was discovered is still called Cornish Avenue.

Gold was discovered in other parts of America and inevitably attracted miners from around the world, with the Cornish always accepted as the leading experts in the field. One of these places was Butte, Montana. The gold proved to be far less extensive than had been hoped and the town looked like dying until it was discovered that there was silver there. That too proved to be of little value. But in 1870, the major discovery was made; there were huge deposits of copper and the site was to earn a reputation as the 'richest hill in the world'. One man who did as much as anyone to build up Butte as a major copper mining centre was a native Pennsylvanian, William Andrews Clark, who not only opened up mines but established the first smelter at Butte. He always gave preference to Cornish workers and it was said that the Cornish foreman at the Mountain View mine, would greet any Irishman who arrived looking for work with the words, 'thee are in the wrong line, my boy!'

Fortunately for the Irish, the other major employer in the region was Marcus Daly, who had left Ireland for the Californian gold fields before arriving in Montana. Mining here continued well into the second half of the twentieth century and there have been a number of plans to restart operations. These have left vast scars on the landscape, but there is one reminder at least of the days when the Cornish were here. Butte is famous in America for its pasties; Joe's pasty shop sells pasties that are said to be at least as good as any made in Cornwall – though one might have difficulty persuading anyone from Cornwall to agree.

Those who came from Cornwall to mine did not necessarily remain in the industry. A.L. Rowse quotes the story of the Rowe brothers. The father had arrived from Cornwall and built a house for himself in Illinois, where he divided his time between farming and small scale mining. His two sons, William and Charles, were among the many lured to the gold mines of Montana. They took the steamer from Omaha up the Missouri to Fort Benton, Montana, and then walked the rest of the way to Helena – today a distance of 132 miles by road. William, the elder of the two, moved around various diggings with some success. He was for a time a sheriff at Deer Lodge, then opened a billiard hall which he later sold and used the money to buy a hotel. He got a contract for carrying mail, bought horses and started a livery business. He ended up having a ranch of 340 acres near Fort Benton, where he raised cattle and bred shire horses. In this busy and colourful career, he found time to get married and raise sixteen children. Charles had an equally busy life, though his mining efforts were not very successful. He got a job driving a stage-coach, then opened a saloon and, like his brother, when that was sold, bought a hotel. When he sold the hotel, he also bought a ranch and would twice become mayor of Fort Benton. Precisely because they prospered, we know their stories; less is known about those who came to mine and never moved on. But the Rowe brothers were certainly not alone in having an adventurous and enterprising spirit that enabled them to thrive.

Wherever there were mines in America, from the gold and silver mines of Colorado to the lead mines of Wisconsin and Illinois, there you would find the Cornish. But America was not the only country in the world to experience a mining boom in the nineteenth century. Australia was also to draw the Cornish to work the mines in vast numbers. But before Australia had its gold rush, it also had a 'copper rush'.

Australia, as is well known, began as a penal colony but, by the 1830s, Australian states were looking to recruit settlers, mainly to work the land. Agents in Cornwall advertised the huge advantages that they claimed emigrants would enjoy, put up posters and gave lectures. They were not specifically looking for miners, as at that date no minerals had been identified. Isaac Latimer, South Australia's agent in Truro, advertised a free passage to Adelaide for married men. He listed the sort of men they were looking for: 'Agricultural Laborers, Shepherds, Blacksmiths, Wheelwrights, Sawyers, Tailors, Shoe-makers, Brick-makers, Builders, and all persons engaged in useful occupations'. Miners were notably not mentioned, but they applied for the free passage anyway. In 1839, 45 of the 360 applicants for free passage were miners, but the following year, there were 132 miners on the list. When they arrived, they found plenty of opportunities to use their skills in jobs such as digging wells and quarrying. But it was inevitable that, sooner or later, a Cornishman would sniff out a mineral deposit. In 1838, James Nicholls came across a vein of silver-lead in the Adelaide Hills and traced it back to the main lode at Glen Osborne. Surprisingly, nothing was done at the time, but three years later, two other Cornishmen, Hutchins and Thomas, found the minerals again in the same area and this time a company was formed and Wheal Gawler had the distinction of being Australia's first metal ore mine. Other companies soon began to form and more mines appeared, but it was the discovery of copper that brought a major mining industry to the region.

Sheep farming was the mainstay of the local economy and it was while rounding up a flock at Kapunda north of Adelaide that Francis Dutton came across a lump of green material that he recognised at once as malachite – copper carbonate hydrate. He mentioned the find to a neighbour, Captain Charles Bagot, who said that his son had also found something similar. That was enough for the two men to acquire the land and begin working the deposit. Their first sample was sent to Swansea for assaying and was found to be 21.5 per cent copper and was valued at £21 per ton. Two mines were begun, Wheal Dutton and Wheal Charles and miners encouraged to come and work at tribute at the handsome rate of 3s 6d in the pound to encourage them. It was not long before news of the find spread. Soon miners were flocking to the Kapunda region, which from an empty landscape soon became a small town. At first things went well, but that old enemy of miners soon appeared – water flooding the workings. The problem was solved by buying a second-hand 30 inch beam engine from Cornwall, which immediately brought about a great surge in production to 2,500 tons per annum, worth over £50,000 a year. Its success, however, was dwarfed by findings by the Burra Burra Creek, right out in the bush, a hundred miles north of Adelaide.

Once again, the find was down to shepherds who brought the news to Adelaide. Two mining companies were formed. One, 'the snobs' was financed by small investors, mainly shopkeepers; the other 'the nobs' was led by wealthy men, including Bagot and Dutton. Governor Grey set a price of £20,000 for the 2,000 acres of land, which was divided between the two groups. For once, it was the little men who came out on top.

The nobs opened up the Princess Royal Mine, and it was successful for a time but by 1851, the capital had all gone and the workings were abandoned. Once again, the land was left to the grazing sheep. The snobs set nine Cornish miners to work under Captain Thomas Roberts in 1845 and success was immediate. The ore was assayed as 71.25 per cent copper, more than three times richer than the Kapunda deposits. Although officially known as Great Wheal Grey, it was soon known simply as the 'Monster Mine', and the Cornish rushed in to the region. The Burra Burra mines were under the direction of the General Superintendent Henry Roach, who made sure that the mine captains he employed were all Cornish. Philip J. Payton in his book *The Cornish Miner in Australia* (1984), studied the records of tributers working at the Burra mines in 1848 when of 63 men, 47 had typically Cornish names and some of the others, with ordinary English names, could also have been Cornish. In 1860-61 135 out of 198 had Cornish names, but this time three Germans were also at work.

A township grew up with a strongly Cornish character. The stone-built houses were similar to those found round mines back in England, and the names looked back to the mens' origins. Although the company town was called Kooringa, the suburbs had names such as Redruth and Lostwithiel, while streets were named after Cornish towns. The men could either enjoy their beer in the Cornish Arms or the Redruth Arms. The success of the mine encouraged the Australians to build up their own infrastructure. Smelting works were developed, one of which was established by Dr Edwin Davy, a relation of Humphry Davy. Where at first they had relied on second-hand engines from Cornwall, they later had the funds to buy new, but there was still the great cost of bringing them half way round the world. In 1852, the Perran Foundry in Cornwall built an 80 inch engine and despatched it to Australia. Once it had arrived, the 40-ton load had to be hauled inland for a hundred miles over unmade roads. The *Mining Journal* described the problems met along the way:

> 'The heavy portions of the monster engine are now on the way up, carried on a huge carriage, specially and ingeniously constructed for that specific purpose and drawn by 72 bullocks. The bridges on the road had to be shored up to bear the enormous weight passing over them.'

Soon foundries were being established, the most successful of which was begun by a Cornishman, James Martin, whose father had a foundry that gave its name to the village of Foundry, now a suburb of Hayle. He emigrated to Australia because he suffered from asthma and felt he would benefit from the hot, dry climate. He established his business and was later joined by Fred May, who had considerable engineering equipment. Soon they were manufacturing all kinds of mining equipment, including steam engines. In later years, they were to become one of the leading locomotive manufacturers in Australia.

The Burra mines prospered throughout the 1840s but suffered a setback in the early 1850s when miners started to leave, lured away by the discovery of gold in New South Wales and Victoria. One could make a decent living working in a copper mine; there was a chance of making a fortune in the goldfields.

The story of gold actually begins in 1823, when an assistant surveyor discovered it near Bathurst in New South Wales, but was told by the government to keep the news to himself. They were worried that if word got out, convicts would abscond and indentured workers leave their masters. The situation changed dramatically in 1849, when instead of a rush to Australian goldfields there was a stampede to California. Now, worried that men might be lured to America, the government actively promoted prospecting and offered a reward to anyone who found substantial, workable gold. Edward Hargraves was not, in fact, a Cornishman, having been born in Hampshire, but he had been out to California, though with very little success. In the country near Bathurst, he found bands of quartz in the hills that strongly reminded him of the gold-bearing lands of California. He met with three Cornishmen – William and James Tom and John Lister – and taught them the rudiments of prospecting and set them to work looking for signs of gold. The enterprise was successful. Gold was found on 12 February 1851 and Hargraves set off for Sydney to report to the Governor and claim his reward. He was awarded £10,000 and appointed Crown Commissioner of Lands. The award was contested by Lister, but the Courts ruled that Hargraves was indeed the man responsible for first finding gold in the region.

The news of the gold discovery soon spread with the predictable result that men crowded into the region. Many were disappointed as the pickings were not as rich as everyone had hoped. However, this turned out to be just a prelude to far greater finds in Victoria, finds that were to make the area the second largest source of gold after California. The scramble to reach the new goldfields far exceeded the rush to New South Wales. Adelaide, which had prospered thanks to the copper boom, was now being deserted as a local reported in February 1852:

> 'What changes have taken place in this colony since Christmas! The discovery of gold has turned our little world upside down, thousand left the settlement for the diggings … In Adelaide windows are bricked up, and outside is written, "Gone to the Diggings". Vessels are crowded with passengers to Melbourne, and the road to the Port is like a fair – ministers, shop-keepers, clerks, councillors, labourers, farmers, lawyers, boys, and even some women, have gone either by sea or land to try their fortunes at the diggings.'

There is no mention in the list of miners, but Cornish miners were to dominate the workings as the goldfields developed. As in California, men had to be tough and rough justice took the place of any legal authority, as one miner, quoted in Payton, reported:

> 'We had to fight for our claim. It was nothing uncommon to find a big rough fellow working your hole and disputing your right to it. This, in the early

> days, had to be decided by a stand up fight, and I must say that it was a quick, if not just, way of arriving at a decision. The diggers would always see fair play.'

In time, the world of individual prospectors gave way to organised mining, with the Cornish dominant. It was said that men looking for work would try and put on a Cornish accent in the hope of convincing the Captain they were genuine, skilled Cousin Jacks. The mines prospered, with output reaching a peak in 1856, when 3,453,744 Troy ounces of gold were produced, roughly 200 pounds.

The mines of Australia brought prosperity to the colony and established strong Cornish influences in the mining regions. The Cousin Jacks brought with them a powerful Methodist tradition on one hand and their own distinct sports, such as Cornish wrestling, on the other. It is no wonder that the copper mining district became known as 'Little Cornwall'.

The third great gold rush took place in South Africa in the Witwatersrand region. As in Australia, the first discoveries were kept secret. In 1852, an Englishman, J.H. Davis, found gold in the region and sold it to the Transvaal treasury for £600, but the government suppressed the news. A fresh discovery was made a year later by Pieter Jacob, but again the rest of the world never heard of it. It was only in March 1886 that an Australian called Harrison discovered gold again, but he was forced to sell his claim for a paltry £20. Harrison was never heard of again, but this time the news leaked out and the Witwatersrand gold rush was under way. Inevitably, the Cornish left for Africa in large numbers. By 1890, the London train from Cornwall had special carriages labelled 'Southampton', the port from which the Cornish would leave for the gold fields. One of the most successful mines was Ferriers Deep that had an almost exclusively Cornish workforce. They were not the first Cornish to make the long trek to South Africa. In 1869, a huge 83.5 carat diamond, known as the Star of Africa, was discovered and began a diamond rush, which also brought Cornish to the region, but it was never quite as important as the gold rush.

The story of the Cornish in South Africa is similar to that of the other miners who headed overseas, but with one important difference. They were not welcomed by the local Boer community, who labelled them *'uitlanders'* – outsiders. They had no voting rights, even if they made the area their home, and the Transvaal government imposed heavy taxes on the industry. The British mine owners bitterly resented the imposition and it was one of the main issues that led to the outbreak of the Second Boer War in 1899.

The Cousin Jacks spread around the world. This was epitomised in a list quoted by Hamilton Jenkins of the fate of nine sons from just one family:

1. 'William' went abroad as a boy of twenty. He paid his first and only visit to his Cornish home when he was a grey-haired man of fifty-five, and then disappeared again for good.
2. 'Dave' died in New Zealand.
3. 'Ki' travelled the world for thirty years and then settled down near his birthplace.
4. 'Jerry' died in Butte City, Montana.
5. 'Wazzy' died in Australia.
6. 'Luther' left as a lad and never came back.
7. 'Johnny' died in Cornwall of African phthisis [lung disease].
8. 'Martin' was killed in New Zealand.
9. 'Willie', called after his eldest brother William because, as his mother said, 'Bill will never come again, and I like the name', visited half the world, made a small fortune, and settled down alone, the only occupant of his father's cottage.

The one item in the list that makes it different from similar lists that could have been compiled for other families, is the final entry. Not many of the Cousin Jacks made even a small fortune. But all of them made a massive contribution in helping developing the mineral resources of the whole world.

Chapter 14

A CHANGING WORLD

The catastrophic fall in the price of copper that followed the development of mines in America exacerbated what was already a worrying situation for the British miners. According to a newspaper account of 1868, production had already fallen in ten years from 147,330 to 88,603 tons a year. Copper had been the mainstay of the whole mining industry in the South West; the value of copper produced in the early years of the nineteenth century was more than twice that of tin. But now there was less and less rich copper ore being found above the 1,000 foot level, and what was discovered was proving increasingly difficult to work. But it was not all bad news. Below the copper ore there was still a rich source of tin and the demand for tin was on the increase. A new commodity was being developed – the tin can.

Tinplate was first introduced in Germany and the process was first developed in Britain in the late seventeenth century, but on a small scale. It became a much more important industry in the early nineteenth century, mainly concentrated in South Wales. Wrought iron bars were rolled into thin sheets, and then had to go through an extensive cleaning process. Scale was removed by pickling in acid and rubbing down with sand and water. The bars were then annealed and again pickled. They were then ready for tinning. The iron was passed through a series of baths of molten tin and grease. The plates were dipped in grease and then in tin, and the process repeated with a final immersion in pure tin. They were then placed in a grease pot to prevent sudden cooling that could crack the layer of tin. With the iron totally covered, the material was unable to rust. At first, use was comparatively limited, being used for a variety of objects from boxes to toys. It received a major boost in the early nineteenth century, thanks to a new process for preserving food.

Napoleon had offered a prize to anyone who could find a way of preserving food, so that it could be supplied to the navy or an army on the move. The prize was won by François Appert, a Paris confectioner, in 1795. His method was to place the food in partially corked bottles which were then immersed in boiling water, then closely sealed. It preserved the food, but bottles are not exactly durable, when carried around on board a tossing ship or by an army heading for battle. The answer was found in Britain by Peter Durand in 1810, who used a tinplate container instead – the original tin can. The idea was followed up by Bryan Donkin, who set up the first canning factory at Bermondsey. They supplied canned meat and soup to the Royal Navy in the American War of 1812. Over the years, canning processes were improved and tinned food would eventually become a commonplace. There was a brisk trade between the mines of Cornwall and the tinplate works of South Wales, especially those of Llanelli, which became known as 'Tinnoppolis'.

Copper production too was to get a boost from a new technology – the development of an electrical industry. The metal with the best conductivity is silver and the next best copper – and there was never any doubt which of the two would find maximum use. The first important invention was the electric telegraph that was developed in a number of different ways by different inventors. The first to be used commercially was a version designed by Cooke and Wheatstone and installed to send messages on the Great Western Railway between Paddington station in London and West Drayton, 13 miles away. It used needles activated by electro-magnetic coils that could point to different letters to spell out a message. The system was soon extended to Slough, where its value received a dramatic illustration in 1845. A murder suspect had been spotted boarding the 7.42 p.m. train from Slough to London. The news was telegraphed through to Paddington and when the suspect arrived at the terminus the police were waiting. The telegraph system was for decades the most efficient way of sending messages over long distances, eventually linking countries around the world. Other developments saw the demand for copper wire and cable steadily increase. Later, the development of electric motors would also have a direct impact on all forms of mining.

The Levant mine on the north Cornish cliffs typifies the changing fortunes of mines in the late nineteenth century. Mining had begun here in the eighteenth century, but the first major workings only began in 1820, when a consortium led by Charles Daubuz and John Batten of the Trereife smelting house took over. At first, the mine was only producing copper but, by 1835, they had moved down to the lower levels of tin. The mine prospered and spread, eventually pushing out over a mile and a half from the cliffs under the sea, with the deepest shaft taking miners down 350 fathoms beneath the sea bed. Wilkie Collins, who we met earlier on his mine exploration, was introduced to this strange submarine world.

> 'After listening for a few moments, a distant unearthly noise becomes faintly audible – a long, low, mysterious moaning, which never changes, which is felt on the ear as well as heard by it – a sound that might proceed from

> some incalculable distance, from some far invisible height – a sound so unlike anything that is heard on the upper ground, in the free air of heaven; so sublimely mournful and still; so ghostly and impressive when listened to in the subterranean recesses of the earth, that we continue instinctively to hold our peace, as if enchanted by it, and think not of communicating to each other the awe and astonishment which it has inspired in us from the very first.'

The noise of the waves, rolling pebbles over the sea bed far over his head may have appealed to the romantic in Collins, but the miners were probably less enchanted. Far out under the ocean, ventilation was a major problem and only continuous pumping kept the sea water at bay, though the underground workings were never totally dry. The value of the ore along this stretch of the Cornish north coast can be gauged by the efforts that had to be made to work the sites at all, as can be dramatically seen at nearby Botallack. The first pumping engine in use, a modest 30 inch beam engine, was installed in an engine house sat on a narrow ledge half way down the cliffs. The effort involved in building with the great blocks of stone must have been immense, not to mention the difficulty of getting the massive engine parts down there and installing them. The space is so cramped that where in most engine houses the stack is outside the main building, here it is inside, tucked into a corner.

The Levant mine was subjected to the changes in the economic climate that affected all the mines in the region. Throughout the 1860s, the fluctuating price of tin caused problems, and these were exacerbated when a dispute arose between the miners and the adventurers. The men claimed that new workings were too close to the sea bed to be safe; the adventurers disagreed but abandoned the mine anyway. In 1871, a new company was formed to reopen the old workings, which required a huge expenditure of £17,500 just to pump out the flooded workings and get everything back in working order. There were times when it scarcely seemed worth the trouble. In 1898, although they sold £10,000 worth of tin, the annual profit was just £146. The mine was developed and modernised by Francis Oats, who had returned to Cornwall after a successful career managing the Beer Diamond Consolidated mines in South Africa. He built a flight of granite steps from the dry at the top of the cliff, with its baths, down the steep side to the top of the shaft and the man engine. They can still be seen, snaking down the precipitous slope. After the tragedy of the collapsed man engine, access to the mine was by an adit close to the sea, which could only be approached down the sides of the deep cleft of Levant Zawn. The mine continued working, producing large quantities of copper and tin to the end, when it failed not just to make a profit but began to lose money. By 1929, the shares that had once been valued at 10 shillings had dropped to sixpence. Levant's working days seemed to be over.

Cornwall in the 1860s was in a depressed state due to the steady fall in the price of copper. In November 1868, the *Cornishman* newspaper reported that the production of copper ore had dropped from 147,330 tons per annum ten years earlier to 88,603 tons. The problem was not merely economic pressures from overseas. The industry had enjoyed boom years and there had seemed to be little incentive to invest in modern machinery. The profits had all been distributed among the adventurers. The mines such as Levant that had moved to tin mining in addition to still mining copper thrived; others simply failed. The result was a great exodus from the county. Many of the mines that failed were simply abandoned, engines sold where a buyer could be found or broken up for scrap. Engine houses were left to decay, though the stout stone walls remain throughout the region as memorials to more prosperous times. Miners' cottages were abandoned, their garden plots going to waste. My wife's family were among those who had lived in Cornwall for generations, traced right back to the sixteenth century, but now they moved out, finding work first at the naval dockyards and later at the Woolwich Arsenal. Others went further afield, joining the Cousin Jacks who had already made their homes overseas. It has been estimated that a third of the mining community left the county and many more were forced to look for different jobs elsewhere in the country. A writer in *Cornish Notes and Queries* wrote at the beginning of the twentieth century, described just what it meant to one family.

> 'My paternal grandfather and grandmother, with their three sisters and three brothers, had a total issue of twenty-six, each of which had a family. My maternal grandparents, with their two sisters and four brothers, had a total issue of twenty-eight, each of which had a family. Of these fifty-four families with a total membership of considerably over two hundred, five only now live in the county … The rest are scattered all over the face of the earth, in London, South Africa, Australia, New Zealand, and America.'

Those who remained relied on a new approach by the adventurers. Too many old practices had been unchanged for decades. This was particularly true on ore dressing, which still relied to a large extent on young women wielding hammers in draughty, open sheds. Below ground, little had changed, with hand work still the order of the day. The nineteenth century had seen several attempts to design power-operated rock drills. An early experiment was made by Richard Trevithick, who designed a machine for quarrying stone for the new breakwater being built at Plymouth in 1812. He described what the machine would do, but frustratingly gave no exact details of the power source, but it would appear to have used a small steam engine, almost certainly one of his portable 'puffers'. He was enthusiastic about its value:

> 'I can bore holes five times as fast with a borer than by a blow or jumping-down in the usual way, and the edge of the boring bit was scarcely worn or injured by grinding against the stone, as might have been expected. I think that the engine that is preparing for this purpose will bore ten holes of 2½ inches in diameter 4 feet deep per hour.'

We do know that an engine was sent to Plymouth and as a result the price of stone fell from 2s 9d per ton to one shilling. Trevithick had obviously learned something from Boulton and Watt; he got a share of the money saved. There is no indication that the system was ever tried in a mine as opposed to a quarry. There were other attempts to make steam-powered drills, but they were never adopted. Improvement depended on a new technology, developed for one of the century's major engineering projects, the Mont Cenis railway tunnel through the Alps.

Originally, work had begun on the tunnel using a steam-powered drill, but Germain Sommelier showed that the same drill could be worked using compressed air. This was an important breakthrough as far as mining was concerned, because if the power plant is based on the surface, a lot of pressure would be lost in the piping between the engine and the drill if steam was used. But there is not the same problem of falling pressure with compressed air. The first successful drill, designed by Low, was manufactured by the Turner company of Ipswich. The original version consisted of a chisel-like bit housed in a cylinder within another cylinder and was made to rotate slightly with each blow of the drill. It was very fast, working at anything from 300 to 500 blows a minute. The first version was mounted on a trolley, but later a lighter version that could be picked up and used by one man was produced. In the new drill, the bit was still housed in a cylinder but was activated by a piston moving backwards and forwards at up to 1,500 times a minute. The chisel was turned slightly at each blow. In the early versions, air was passed through a tube to blow out the fragments produced by drilling. The clouds of minute rock particles made work difficult and the early attempt at a remedy was simply to spray water round the drilling site to keep down the dust.

In 1881, John Henry and James Miners Holman had taken over the family business from their father. One of their first actions was to sign an agreement with James McCulloch, who had invented a new type of rock drill, and they took out a joint patent. It became known as the Cornish Rock Drill and by 1882 it was in use at Dolcoath, South Crofty, Tincroft and East Pool. It was an immense success and by the 1890s, the drills were being exported around the world – there were said to be 1,000 in use in South Africa, most of them supplied by Holman's. In 1910, the South African Chamber of Commerce organised a world rock drilling competition, in which Holman took first and third prizes, establishing them as one of the leading manufacturers. In order to use the new drills, the company needed compressors. These were powered by steam and in the early two-cylinder versions, the steam moved the piston in the first cylinder and was connected by a rod to work the second piston in the compression cylinder. Later versions used two pairs of cylinders, set either side of the flywheel. The first machines stored air at 60psi, later versions increased the air pressure to 80psi. The compressed air was stored in a steel tank, ready for use.

They installed the biggest compressor used in Cornwall at that time at Carn Brea in 1894. All that remains on site now is the compressor house chimney. But there is no doubt that the introduction of the rock drills was a key factor in enabling Cornish mines to continue operating into the twentieth century. Holman continued in business as an individual company until 1968 when it merged with Broomwade to form Compair Holman. In 2003, the Camborne works were closed, ending two centuries of invaluable work for the local mining industry. The closure was sad news and just before the end came, when the staff were clearing out cupboards, they came upon a remarkable hoard of film footage, some of it dating back to the early years of the twentieth century – a travelling shot, that made use of the overhead gantry in the factory and showed a major industrial complex with a large workforce. I presented a short film based on the old Holman footage for BBC *Inside Out*, and this invaluable material is now being conserved.

The mines that kept going through the second half of the nineteenth century faced a difficult period in the nineties, when the price of tin slumped from £100 a ton at the start of the decade to £64 in 1894. A number of famous mines closed down and, once pumping stopped, the works would be flooded out and require a great deal of money to reopen, even when the metal prices rose.

There were some who saw the closures as being as much due to the slackness of the mining community as international ore prices. A Cornish mine manager described a working day as he saw it in 1896:

> 'They start to go down at 6 a.m. the last do not reach the bottom before 6.45. Then they wait for their tools to be sent down, and about 7 o'clock are getting to their places. Before working, however, they have "croust". I never saw men in Levant or in any mine in America or Africa eat underground before working, but here it seems to be the custom. After that they begin to work and keep it up till 10.30. Then they have more "croust", and resume work about 11.30. After another 2½ hours at it they make their way towards the shaft and are ready for the first turn of the gig.'

This means, according to the manager, that they only worked 4½ hours a day, which sounds unreasonable. But if one looks a little closer, things start to look rather different. For the men to arrive at work at that time they may have had to leave home at 5.30 or even earlier and one can hardly be surprised that they were feeling hungry since they probably had no time for breakfast. The time taken in going down the mine and making often long journeys on foot to and from the actual work station do not apparently count as work. Then, for most men, the work was still all manual labour as this was before the pneumatic drills came into operation. In many of the old mines, ventilation was bad and temperatures could reach 32°C. In those conditions, only the fittest could manage more than these men did.

In spite of the circumstances, copper mines did flourish for a while in the second half of the nineteenth century. In Cornwall, the mines of Caradon Hill with the prosperous South Caradon and Phoenix United mines, were doing particularly well. Between 1843 and the end of the century, their tin and copper sales were worth over £1,300,000. The mines were, in fact, so successful that they acquired their own transport route, the Liskeard and Caradon Railway, which opened in 1844. This was a typical tramway of the period, with rails mounted on stone blocks, many of which can still be seen snaking round the hill. The tracks took laden trucks down to Moorswater, where they were loaded into barges on the Liskeard and Looe Canal. The outward journey from the mines was all downhill, the trucks moving under gravity, controlled by a brakeman. The returned empties were hauled back by horses. The line was so successful that the owners of the Liskeard and Looe Canal built a railway parallel to the canal down to the quays at Looe and hired a locomotive. The line opened for business in December 1860.

In Devon, the greatest mine was based on the discovery of copper in a stretch of woodland called the Blanchdown Plantation. Unfortunately for the adventurers keen to exploit the riches, the woods were owned by the sixth Duke of Bedford and were used for raising pheasants. To the Duke's way of thinking, shooting birds was more important than mining. The seventh Duke, however, was more interested in money than shooting, and in 1844 he agreed a twenty-one year lease with local mining engineer Josiah Hugo Hitchens in exchange for one twelfth the value of all the ore raised.

The mine at once proved so prosperous that the men could not move the ore quickly enough.

> 'So sudden and unexpected had been the discovery and in such quantities was the ore being thrown up that the place was crammed with copper, the richest yellow sulphuret in appearance and much more resembling heaps of gold than the baser metal and representing a combination of gorgeous metallic wealth and sylvan beauty the like of which will never be seen again.'

The mine continued to prosper greatly and in 1868, the Duke's mineral agent wryly noted that if he had chosen to develop the mineral rights himself, the Duke would have been pocketing more than a million pounds. On the 21st anniversary of the opening of the mine, a huge feast was laid on for every man, woman and child who worked the site. Some idea of the numbers involved can be gauged by the food and drink supplied – a ton of beef, 1,300 pounds of bread, eight hogsheads of beer (432 gallons), four of cider and just one of lemonade. Captain Isaac Richards made a speech in which he reported that 420,000 tons of ore had been raised and sold and underground workings now stretched for 25 miles. But it was not to last; the returns began to diminish and, like other copper mines, falling prices steadily reduced profitability. A saviour appeared, however, in the form of an unpleasant beetle that was causing havoc across the cotton plantations of the American South – the boll weevil. Not that everyone hated the beast. In Centre, Alabama, there is a strange statue of a lady in a Grecian type robe holding a dish above her head on which perches the giant beetle. It appears that the local farmers abandoned cotton in the face of the devastation, and turned to growing peanuts – and made far more money than ever before – hence what must be the only statue in the world to a predatory beetle. But for most of the cotton farmers, it was a disaster and the only solution seemed to be spraying the crops with an arsenic-based insecticide. Devon Great Consols and other copper mines knew there were considerable quantities of arsenic in the lode and began exploiting it.

The arsenic was found both underground and in the dumps of low grade ore that had not seemed worth smelting. The ore was broken up into pieces small enough to pass through a ½ inch mesh sieve. It was then calcined, and the vaporised arsenic passed through flues with thick masonry walls to a complex of brick chambers where most of it condensed. The remaining vapour was passed over limestone screens to remove as much of the remaining arsenic as possible. The vapour was then discharged up a tall chimney. The calcining was then repeated, when the pure arsenic was deposited in chambers with tiled floors. The job of removing the white crystals was all carried out by hand by men wearing improvised face masks and ear plugs and with feet wrapped in sacking to prevent what was known as 'arsenic sores'. It was an unsavoury job, with severe health risks, but the enterprise was profitable. Production began in 1866 and in the first two years, 437 tons had been sold.

Not everything went smoothly at the mines. There were problems between the workforce and the owners, initially caused by them employing unskilled workers at lower wages. This was bad news for men working on tribute, as they relied on efficiency from everyone working with them – the unskilled slowed everything down and when things slowed down, the tributers' pay went down as well. The men got together and formed the Miner's Mutual Benefit Association – a trade union in all but name. One demand was that a committee of miners should be formed to assess whether or not the price being offered for a particular pitch was fair or not. To the owners and agents this was heresy – the men had no right to have any say in their wages. The response was immediate – any miner joining the Association was locked out.

Real trouble erupted in February 1866, when a blackleg who had been working at the Drakewalls mine near Gunnislake was grabbed when he came off shift and mounted on a pole. He was then run through the streets of the town to loud cheers. The mine manager was alarmed for his own safety and armed himself with a knife and when that was taken from him, he called in the police. The result was that the Drakewalls men found themselves carted away and imprisoned. The arrest and conviction of the men caused enormous anger, especially in Tavistock, where there were many Association members. Threatening letters were written and there was a real danger of direct action that might have included stopping the pumps, resulting in a mine being flooded out – incurring a huge cost to reopen it later. The owners arranged for 150 constables to stand guard at the mine with 150 special constables, recruited for the occasion, to reinforce them. Even that seemed too little for the owners and 150 troops arrived at Tavistock from Plymouth. It seemed as if the area was under siege.

An immense crowd gathered at the Devon Great Consols count house for the usual auction of pitches for tribute and tut work and the owners were clearly worried about what would happen. Three hundred police were drafted in to oversee the procedure. The work was being offered at what the men considered ludicrously low prices and, as a result, only four bargains were struck. Attempts to persuade the men to take up the work were shouted down and the meeting broke up without resolution. A similar scene was enacted at the Hingston Down mine near Gunnislake and once again the mine owners were able to call in the military – a government launch brought a detachment of armed soldiers to Cotehele. The miners were desperate but the times were against them. They might stay away from the mine, but that did not mean that work would come to a standstill. This was the period when mines were being closed down in west Cornwall, and there was no problem finding men willing to work the mines of the east. The Association collapsed and men had to return to work at lower wages.

The other great bone of contention was the five-week month. When accounts were settled, it was on the basis that a month consisted of five weeks – and that was how long the men had to work to get their month's pay, which in effect meant working some days for nothing. When the Association had crumbled, the last deal they had agreed to was for a reduction in the weekly rate but on the basis of a four-week month. Then, in 1866, an attempt was made to reintroduce the five-week month, but the move was successfully resisted – a rare victory for the miners and the bells of Tavistock rang in celebration. But the success was short-lived. The mines were losing money and in 1879 Devon Great Consols was said to be making a substantial loss. The owners reacted by trying to cut wages and an attempt to strike failed in the face of company opposition. The great days of the mine were almost over; for a time, profits came almost entirely from arsenic and numerous attempts were made to try and find tin at the deep levels below the copper lode. But by the end of the nineteenth century, it was clear that the mine was failing and in November 1901, 351 men were laid off and a handful kept on to keep the pumps going in the hope of better times ahead. One of the problems that beset the mine was a failure to modernise – and the world of mining was changing rapidly, even if many owners were unable or unwilling to embrace the new technology. In the next chapter, we shall be looking at one mine that was developed as state of the art for the beginning of the twentieth century.

Chapter 15

THE CAMBORNE SCHOOL OF MINES

Throughout the early years of the mining industry, technical education had scarcely been available; one learned on the job. But in a world where technology was changing at a rapid rate, it was evident that something more was needed. The Miners' Association of Cornwall and Devon was formed in the 1850s and began running a series of evening courses. In 1863, they became formally attached to the National Science Art Department, which enabled them to give students officially recognised certificates on successfully completing their courses. The result was the establishment of Science and Art schools in Camborne, Redruth and Penzance. In 1888, Camborne also established the Mining School, which would soon become the principal mine school for the whole area and would achieve worldwide recognition for its teaching.

In the classrooms, there were lectures on the theory and practice of mining, covering all aspects of the industry. But mining is a very practical activity. At first, the practical side relied on sending students to work at various mines in the area, but it soon became apparent that what was really needed was a mine which was under the control of the school and where specific tasks, from underground surveying to learning about explosives, could be taught in a controlled environment. What they needed was a mine that was not too deep, so that students didn't have to spend half their time going to the working site and one that did not need pumping, as that was one of the most costly and time-consuming activities. In 1897, agreement was reached to take over the South Condurrow mine, between Camborne and the village of Troon. Work had stopped there in 1895 due to the low price being paid for tin. The main source of ore was the Great Flat Lode, which was also worked by the nearby Wheal Grenville, but at a lower level. The Wheal Grenville workings had joined in with those of South Condurrow, as a result of which any water from the latter was removed by the Wheal Grenville pumps. Once the mine was acquired, it was decided to work it for tin, using a small staff of experienced miners and a far larger number of students to help them. Any money raised from the sale of tin would help pay the expenses of running the school. A very full account of how the mine was developed and used over the years can be found in Tony Brooks and John Watton's *King Edward Mine* (2002), from which most of the following information has been taken. What makes the story of this mine so important is that, as it was being totally refurbished, everything that was being installed represented the latest ideas on mining technology.

There was an immense amount of work to be done on site. There was one whim engine, served by a somewhat antiquated boiler. The engine house over what had been the Engine Shaft had already gone. There were no ore dressing facilities left at the surface. One building that had survived was the miners' dry, which was refurbished in a way that made it seem like four-star hotel luxury compared to the facilities available at many other mines. Heat was supplied by a boiler tube running the length of the main room. It also provided hot water for eight or nine plunge baths and two showers. This facility was for the students, but a similar arrangement was made for the professional miners. A new building was erected to house a compressor that could drive two rock drills. The next new building was the single-storey survey office, where all the surveying lessons were given.

A new headgear was erected at the Engine shaft – a complex business that used shearlegs. When in place, the centre of the pulleys was 44 ft above ground level. The cage was attached to the hoisting rope with Humble's patent hooks. This was a device to prevent over-winding, which can occur if the cage rises too high above its proper stopping point and that can result in the hoisting rope snapping, sending the cage and anyone unfortunate enough to be in it hurtling down the shaft. The mechanism effectively stops the cage moving up too high by detaching it from the rope. A photograph shows the cage in use with three men rather squashed together inside it and only a single bar across the front for protection.

By the end of the century, the mine had a new name – King Edward Mine, the name it has retained right up to the present day. The big difference in underground working was the use of powered drills. The photograph shows a heavy drill being used, which is rested on a stand. The drill is moved forward by a screw feed, operated by the miner's left hand. The other young man in the picture is holding a water spray, fed from the pressure tank seen in the foreground. This was supposed to keep down the potentially lethal dust that flew out of the hole as the drill bit into the rock. Black powder had now been replaced in blasting by gelignite. This required a different technique from that used for gunpowder that will be described in more detail in the next chapter.

The whole mine was organised in a very efficient way. Once the blast dust had settled, the men moved in and removed any loose rock, before the ore was removed and lowered down to be loaded into wagons. The wagon would then be taken along the level to the bottom of the shaft, loaded into the cage and brought to the surface. From the headgear, the wagon would be taken on an elevated tramway to be tipped into an ore feed above the stamps, ready for processing. It was here that some of the most important changes were being made. The first stage was simply to hold back the larger chunks behind iron bars that formed a crude sieve known as a grizzly. From there the smaller pieces went straight into the ore bin, while larger chunks were first passed through the jaw crusher.

The crushing of ore was important and careful regulation was needed if all the tin was to be extracted. If the particles were too fine, they could easily be washed away in the separation process. In Cornwall, steam-powered stamps had been in use for a long time, but there had been few attempts to improve their efficiency. In the mining booms of the late nineteenth century, particularly in America, numerous improvements had been made, but had generally not made their way across the Atlantic. A set of California stamps had been exhibited at the Paris exhibition of 1900 and this set was bought from the Chicago engineering company Chalmers & Fraser for the mine. The new stamps were very efficient. Each of the five stamps had a replaceable shoe that was wedged into the bottom of the stamp, held in place by scrap wood wedges. The stamps were worked by cams and the heads rotated slightly at each stroke to ensure an even wear on the heads. The ore was fed into the mortar box below the stamps, where it was mixed with a steady stream of water. One of the unique features of the California stamps was the way in which the ore was automatically fed into the box, regulated by the up and down movement of the stamps themselves. If there was a long drop before the head reached the ore, then more could be fed in; if it was scarcely moving, then the flow of ore would slow down or stop.

The Holman steam engine that was to power the mill was unlike the familiar beam engines that usually powered the Cornish stamps. This was a horizontal engine built by Holman with power transmitted by belts, a system that had been used for some time in textile mills, but was uncommon in the mining district. The mill building itself was newly built.

The material from the stamps consisted of a little free cassiterite, rather more cassiterite mixed with waste and – the bulk of the material – just waste. The next task was to separate out the valuable ore from the waste and the principles had not changed; separation still depended on the fact that cassiterite is heavier than the other materials. Gravity would help with separation. But it was recognised that separation could be carried out more efficiently if the material was first sized; the crushed ore being classified as rough sands, fine sands and slime, by passing through a hydraulic classifier. The larger rough sands were treated on a Buss table. This was another example of using the latest technology – the patent application had only been entered in 1899. The system was very similar to that described earlier. This too was a vibrating table, in which the heavier, richer tin was caught near the top of the table and the rest spread out. The high grade was removed and the middling material sent for pulverising. The waste was washed out into the tailings launder and was eventually washed away into the river.

The middle products were ground in a barrel pulveriser, a bit like the drum of a modern washing machine, except that instead of being filled with washing powder, this drum contained steel balls to grind down the material. Once again this was modern technology; the machine was patented in 1880. After pulverisation, the material was resized and the fine sands were passed to the Frue vanner. The first Frue vanner in Cornwall was installed at Dolcoath in 1898, and proved very successful, so was quickly adopted at King Edward. The machine consists of an endless rubber belt mounted on the frame and set at a slight slope. The belt moves slowly along the frame, up the slope, but is also given a slight side to side movement as well. The material is added at the top of the slope together with a stream of water. The sand is kept in a constant state of agitation, and as well as the water that is carried over with the sand, jets of water are added at the side of the belt at regular intervals. The result is that the heavy cassiterite clings to the belt and is carried up with it, while the looser waste is washed away.

The slimes were treated with an Acme table, an improved version of a device that had been in use for some time. Essentially, it consisted of a concave wooden table that was slowly rotated. The slime was introduced gently down a curved launder that ran round two thirds of the table. As the slime slowly moved down the table, the heavier material tended to stay at the top of the table, where it could be brushed off, while the waste dropped to the centre and was washed away.

The final addition was a calciner. There was a certain amount of copper mixed in with the tin. Once heated in a reverberatory furnace, the copper could be extracted by dissolving it in dilute acid. The metal could then be precipitated using scrap iron, an early ion exchange process. There had been some criticism of the set up at King Edward, though the criticism was scathingly dismissed by the London and West Country Chamber of Mines:

> 'Criticism of this description generally emanates from those who, knowing nothing of mining, know less of technical education'.

The organisation pointed out that although the machinery might look like a mish mash of different types, it was designed to give the students a chance to work with as many different types of equipment as possible, and then went on to list everything that had been installed:

> 'Dealing only with the dressing of the ores they have a Blake stone-breaker; Fraser and Chalmers' Californian stamps; Challenge ore-feeder and Frue Vanner; three Luhrig classifiers; one Bartle's improved pulveriser, latest type, working on a roller bearings and belt driven from centre of barrel; one Acme table (Holman's patent); and one Buss table and several automatic samplers… That it is the perfection of dressing is not contended, but that it is calculated to help the student towards the determination of that important problem is beyond question.'

The School of Mines continued offering a specialist three year diploma in mining engineering and alumni could be found working at mines across the world. In the 1970s, however, it was decided to broaden the curriculum and to offer degree courses. The old facilities were inadequate, so a brand new purpose-built facility was opened between Camborne and Redruth. Courses were no longer taught at King Edward and the mill machinery was stripped out and moved to the new site. It was not, however, the end of King Edward. A group of volunteers began planning to restore the mill and its machinery as a museum of the industry. It was not possible to recover the original machinery, but this was a time when the industry was in its death throes and redundant machinery was available and could be installed and put into working order. On some open days, visitors can enjoy the dubious privilege of being deafened by the sound of the California stamps at work. Today, the School has been incorporated into the University of Exeter and in 2004 the whole operation was moved to the Tremough campus, Penryn. Throughout its operation, the mine was adapted to changing circumstances and the new industrial developments of the twentieth century.

Chapter 16

THE LONG DECLINE

Technological advances sometimes increase productivity, but at a human cost. This was very much so in the early years following the introduction of the compressed air rock drills. A Parliamentary Report on the health of Cornish miners, published in 1904, produced some horrendous statistics. Lung disease caused by conditions down the mines had been a scourge for centuries, but seldom affected men under the age of forty. Now the situation had changed.

> 'During the last few years, however, there has been an enormous increase in the death-rate from lung diseases, particularly among younger men from about twenty-five - forty-five is now far greater than at any previous period during the last fifty years. Between the ages of twenty-five and forty-five the death-rate from lung disease among miners living in Cornwall has recently been from eight to ten times the corresponding death-rate among coal miners and ironstone miners.'

The explanation was not difficult to find. The rise in the death-rate only occurred in men working with the new drills and did not occur among those who were still using hand drills. As the Commissioners noted, the men working an end where machine drills were being used were soon covered by dust, 'and the grit is soon noticeable in the mouth'. According to the Report, the damage to the lungs made the men particularly prone to contracting tuberculosis. The problem was exacerbated by Cornish miners returning from South Africa, where in the dry mines the men emerged coated from head to toe in fine white stone dust. Some of these men may have brought the tuberculosis bacterium with them and contributed to the spread of the disease. There is an alternative explanation, that the disease was misdiagnosed and the men were actually suffering from a form of silicosis. But whatever the diagnosis, the result was deadly and an answer had to be found. At first, it was thought that simply spraying water at the drilling site, as was done in the early years at King Edward, would keep the dust down, but eventually, as mentioned earlier, a more successful solution was to inject water at the point of drilling. Having recognised the damage caused by fine rock dust, other steps were recommended to ensure the men's safety when shots were fired. Fans were introduced to clear the air and the men were forbidden to enter the end before the dust had been removed. The Commission ended their report on an optimistic note: 'there is no reason why work underground should not be a perfectly healthy employment, being wholesome to body and mind'. Anyone who has visited a working mine might question this somewhat Panglossian view of life underground.

One change at least was wholly beneficial; the introduction of cages to take men up and down the main shafts. They could be very modest affairs, such as that at King Edward Mine, described in the last chapter, where three was a crowd. Throughout the latter years of the nineteenth century, these would all have been worked by steam. Where they existed, they ended the days of climbing the ladders each day and were arguably the greatest improvement in the welfare of the miners since deep mining was first begun.

The life of the country as a whole changed dramatically in 1914, with the outbreak of the war. It soon settled down to a war of attrition and the horrors of trench warfare. An alternative to trying to attack the enemy by advancing across no-man's land was to attack his trenches from underneath, literally undermining them. Digging tunnels from the British trenches to the German lines was clearly a job for which the mining community was uniquely suited and recruiting officers toured all the mining districts of Britain; eventually, some 40,000 miners signed up and thirty tunnelling companies were formed. They may have been affected by the almost hysterical wave of patriotism that was engulfing the country, but many may also have been tempted by the high pay of 6 shillings a day. In 1915, the recruitment officers arrived in Hayle and the result was the formation of the 251st Tunnelling Company of the Royal Engineers. They arrived in the Loos area in October 1915 and remained in France until 10 August 1917, when they fired the very last mine to be set off by the Royal Engineers.

A typical tunnel was just four feet wide and three feet high, worked out from a shaft sunk to a depth of twenty to thirty feet. Very few men could be set to work at any one time in the cramped conditions. Some would be working away at the face, while their colleagues would be hauling away the spoil. It was necessary to work as quietly as possible – if discovered they might well be the ones blown up instead of the enemy. The Germans were conducting a similar campaign. There was always the possibility of meeting a tunnel coming the other way. The following story from those days was not untypical.

Lt. Geoffrey Boothby was in charge of a tunnelling squad, although he had absolutely no experience of mining, having recently completed a medical degree. One day, he led a party to inspect an old tunnel in which someone had found a German telephone. It appeared that the Germans had been using it to drain one of their own tunnels, so the British soldiers had to slosh along noisily, ankle deep in water. They were approaching a point where the tunnel turned through a right angle, when a gun appeared around the corner and started firing. The German rifleman had no need to show his face or try to aim – in that constricted space any bullet had a good chance of hitting someone. In the event, they all escaped, but it had been a close thing. Later they returned and blew up the tunnel. Boothby's luck ran out a year later, when a bomb he was inspecting exploded – his body was never recovered.

The work of the tunnellers could be used to devastating effect, and never more dramatically than at the Battle of Messine Ridge in France in 1917. The Germans had occupied the high ground ever since the Western Front had been established, and there seemed no way of attacking them directly. In 1916, work began on the tunnels, twenty-two of them, each one of which ended directly below the German trenches and was packed with high explosives, 450 tons in all. The charges were to be set off simultaneously on 7 June. General Plummer, the officer in charge, remarked laconically, 'Gentlemen, we may not make history tomorrow, but we shall certainly change the geography.' On the day of the explosions, the artillery set up a heavy bombardment of the German positions, the usual prelude to an attack by ground forces. This sent the German soldiers rushing to their defensive positions to repel the expected advance. Twenty minutes after the barrage had started, the explosives were detonated and a vast column of earth shot into the sky, estimated to be as high as the dome of St. Paul's. Around 10,000 Germans were killed more or less instantly, and the remainder were so shocked and deafened by the explosion that they offered no resistance as the British tanks rolled in. Messine Ridge was taken, and the ridge itself changed for ever – today the craters are a string of ornamental lakes.

Two months later, the last deep mine of the war was exploded at Givenchy-en-Gobelle near Arras, the mine at the northern end of the sector was exploded in a tunnel dug by the Cornishmen of the 251st Tunnelling Company. Over the years, they had surely earned that extra pay, both in terms of the dangers they faced and all too often failed to survive, but also for the deadly effectiveness of their work.

The twentieth century was to see huge changes in technology. One of the simplest was the replacement of the candles stuck on the miners' hats with more efficient lighting systems. In the 1890s, a process was developed for producing calcium carbide commercially. If water is added to the compound there is an immediate reaction, producing acetylene gas and leaving behind slaked lime, calcium hydroxide. The acetylene burns with a bright flame and by 1900 was being used in lamps, originally it seems for mounting on bicycles. Soon miners' lamps were being developed. The carbide was housed in a lower chamber with the water in a reservoir above it. The drip of water was controlled by a valve that could be operated to regulate the flow and brightness of the flame. In the mines, the carbide would be used up before the shift ended, so the miners would carry a tin of carbide; the slaked lime would be scraped out and the new material added. These lamps were far more effective than the candles and remained in use until the electric battery lamp was introduced; the lamp that is still in use in mines around the world to this day.

Underground, one of the biggest changes came to Cornwall, as it had in other parts of the world, at the end of the nineteenth century with the use of new high explosives, based on nitro-glycerine. Where black powder worked because it burned extremely quickly, creating a heaving force that broke the rock, nitro-glycerine-based explosives detonated, sending shock waves through the surrounding rock, creating fractures which were then blown apart by the resulting high-pressure gas, making for a far more powerful explosion. This was good news for the miners, as the same effect could be achieved by using a small number of holes packed with high explosive instead of having to bore a lot of holes to pack with gunpowder to produce the same effect. There was a further advantage in that the fumes from nitro-glycerine were less noxious than from gunpowder, though the former could produce both carbon monoxide and other fumes. Nitro-glycerine is an oily liquid and is very unstable. Stabilising the explosive was first achieved by Alfred Nobel in 1867, who combined the nitro-glycerine with the earth kieselguhr to create dynamite. In 1873, Nobel developed blasting gelatin, which was created by dissolving colloidal cotton in nitro-glycerine with the addition of wood pulp and saltpetre. Unlike black powder, the new explosives could not simply be lit using a flame, but needed some sort of detonator. In the early days, the detonator cap was a copper tube filled with two substances, mercury fulminate and potassium chlorate, which was attached to the end of the safety fuse and then buried in the gelignite.

Gunpowder manufacture had begun in Devon in 1844, when a wealthy Plymouth businessman set up a powder mill in the heart of Dartmoor. The site was well supplied with water for power and had the great advantage that it was remote from any habitation. There was an obvious good reason for this as, even when the most stringent precautions are taken, there is always the risk of a devastating accident. Gunpowder mills are either built of sturdy stone to withstand any blasts or of flimsy planks that can't cause a lot of damage and can easily be replaced. As the move to high explosives got under way, two factories were established in Cornwall. The first was the National Explosives Company, established in the sand dunes at Upton Towans to the east of Hayle. It was opened in 1888 and by 1890 was producing 5 tons of nitro-glycerine a day. Production increased in the First World War, when the workforce was over 1,800, mostly women, and the works turned out 2,000 tons of explosive, mainly cordite, for the forces. It closed in 1919 though it continued to be used as an explosives store until the 1960s, when it was demolished. The site today has a rather more peaceful use, as a large part of it is now home to a caravan and chalet park. This enterprise was followed in 1892 by the British and Colonial Explosives Company with works at Cligga Head near Perranporth, which was later bought by the Nobel Company. A slump in metal prices in 1905 led to the closure of the plant, but it was reopened at the outbreak of war, only to close again for good when it was over. The same site was also known as Cligga Mine, a wolframite mine – the mineral that was smelted to produce wolfram, also known as tungsten. Tin was also mined at Cligga. It is strange to think that an industry that once employed nearly 3,000 people has left so little trace on the landscape.

Improved drills and explosives increased efficiency, which needed to be matched by more efficient ways of getting the ore from the mine to the surface. The appearance of high explosives led to the introduction of the grizzly to hold back the bigger blocks that could have jammed up the system. When a load of heavy material had accumulated, it was covered by mud and other material and shattered by exploding with dynamite. The ore from different levels was taken to passes, where it was then passed through the grizzly, all of it eventually arriving at the lowest level and the loading pocket at the foot of the shaft for raising to the surface. The system could not be used earlier as gunpowder alone would not have broken up the big blocks.

Hoisting had traditionally involved using kibbles, rather like oversized, rounded iron buckets, originally raised by hemp ropes, but later by chains. Very few shafts were vertical, so that the kibbles were dragged along timber-lined shafts. This was satisfactory up to a point, but it only needed one link to fail for the whole to come crashing down the shaft, causing great damage and possible injury. The system was improved in the latter part of the nineteenth century with the introduction of wire ropes. But even so, it was a far from perfect system. At the same time as the wire ropes were introduced, skips came into use. These were sheet metal containers, fitted generally with two pairs of wheels to each side, that ran between wooden guide rails in the shaft, which would generally be inclined. One could think of it as a railway turned up on end. When the skip reached a pass, the ore was simply tipped in through a chute. In hauling, the rope was kept in place by means of sets of rollers. When it reached the surface, the shaft was closed behind the skip, and the lander knocked out a retaining pin that opened up the bottom of the skip allowing the ore to pour out. In later developments, two skips were used, one counterbalancing the other to some extent. This system was greatly improved later in the twentieth century.

Throughout the nineteenth century, the steam engine had been the main source of power for the industry and remained so for an astonishingly long time and none lasted longer than the engine that was originally built by the Copperhouse Foundry in 1854 for the Alfred Consols mine near Hayle. This was an 80-inch pumping engine that was said at the time 'not to be excelled in the country for duty and appearance'. The engine was moved around various mines, with spells working at Wheal Abraham and then went to Wheal Vor, after being fitted with a new cylinder by Harvey's. In July 1885, there was a fire at the engine house that started in the early hours of the morning and which the local fire brigades were unable to contain. It was assumed that the blaze was so fierce that the old engine would have to be scrapped, but with a few new parts it was soon back at work. In 1903, it was moved to the Robinson shaft at South Crofty, the massive beam being towed to the site by a pair of traction engines, The engine remained in use there and by the middle of the twentieth century was still delivering 310 gallons a minute from a depth of 337 fathoms at the rate of three strokes per minute. The engine finally stopped work after more than a century in use in 1955.

The Robinson 80-inch was a beam engine, as were all the early engines, whether being used for pumping, winding or working stamps. But long before she stopped work, more efficient horizontal winding engines had been introduced. These were, as the name suggests, engines with horizontal cylinders, in which the piston was attached to a piston rod that joined it to a crank that turned the drum. The engines could be geared and many drove twin drums. They worked smoothly and efficiently and were faster than beam engines of similar size. Such an engine was installed by Holman Brothers at the King Edward mine in 1908 and was taken to the wolfram mine at Castle-an-Dinas where it continued in use until 1957. The engine has now been restored and is back at King Edward, where it is regularly demonstrated using compressed air. But by the time the engine was first installed, other power sources were available.

The internal combustion engine evolved during the nineteenth century, first with the gas engine then with the diesel or oil engine. The latter could be used for a variety of purposes, powering compressors or pumps, for example. The advantages were obvious. Steam engines required coal to feed the furnaces, which had to be imported at considerable expense, mainly from South Wales. Oil could be brought in at a far lower cost. With the internal combustion engine, there was no longer any need to employ a stoker and they did not require a special kind of sturdy building as housing. They were never, however, very widely used, for early in the twentieth century, a much more useful power source became available. In 1832, Michael Faraday had demonstrated both how electricity could be generated by the movement of a conductor in a magnetic field and the converse, how an electric current could be used to create movement. At first, no one seemed to be prepared to develop these ideas on an industrial scale, and many sceptics simply regarded the whole thing as an amusing game. When a leading politician enquired what use this electricity might be, Faraday gave him an answer that would have gladdened any politician's heart – one day you may be able to tax it. By the end of the century, however, the value of electric motors was widely accepted.

The first electric generating station was established in St. Austell to serve the china clay industry in 1886. It was 1899 before the mining industry decided to build a generator. The earliest proposal was for a water-powered generator, but in 1899, work began on a more ambitious project, with the construction of a power station at Carn Brea. Four Bellis and Morcom steam generators were installed of 425 kW capacity that also served Camborne, Redruth and Illogen to a limited extent. It was the opening of the Hayle power station in 1909 that brought power to the wider area. Others were to follow and the mining industry began the long process of transforming from steam to electric power. The early generating stations generally used either gas or oil engines to provide the power for the DC dynamos, The first electric pumps were installed at the Tywarnhaile Mine near Porthtowan. These were turbine pumps, consisting of a series of discs – impellers – in a suitable casing. Water was fed into the centre of the first spinning disc and thrown out by centrifugal force into a diffuser that converted the energy into pressure, before the water moved on to the next disc in the line and so on until enough pressure had been obtained to lift the water from the mine to the surface. Electric ram pumps were also used. Soon electric motors were doing all the jobs that once had been carried out by steam.

The big change in tin dressing came early in the twentieth century, with the introduction of California stamps with automatic feeders, classification, vanners and shaking tables. The whole system was automated and the old system, based on hard, manual labour, was ended. The other important change in ore dressing was the introduction of the flotation process for separating out the unwanted iron, arsenic and copper sulphides in the tin concentrate, jobs that had previously relied on calciners. This method depends on making some mineral surfaces repel water. The process is usually improved by the addition of chemicals that attach to the unwanted minerals to ensure they are water repellent. When air is blown through the pulp, the water repellent material attaches to the air bubbles and floats to the surface where it can be skimmed off. The process only worked for sulphides but, in the 1970s, a flotation system was installed at Wheal Jane in an attempt to recover very fine tin. Among the mining community, the greatest change undoubtedly came with the gradual abandonment of the old tut and tribute system in favour of the simpler wage system for some workers, though many worked to contracts, still being paid in terms of what they produced rather than the hours they put in.

The industry as a whole endured something of a roller coaster ride in the first half of the twentieth century. At the beginning, there was a marked effect, known as the 'electric boom', when operating costs could be reduced by switching from steam power to electric motors and pumps. Not everyone embraced the new technology; some of the smaller mines lacked the necessary capital, and some other owners preferred increased dividends in the present to investing for the future. Nevertheless, these were reasonably good times for the industry as a whole, in spite of ever growing output from mines in the Far East. Many small mines were started up, though few survived for long. The outbreak of war in 1914 had inevitably brought more changes. It was not simply that men were recruited into the forces, but the whole ethos of mining changed. The demand for metals increased, and not merely for tin. Tungsten was used as an alloy with steel, making the metal harder and more resistant to wear. Between 1913 and 1916, the output of tungsten doubled. Castle an Dinas opened at this time – probably the first mine to be opened on a previously unknown lode for generations. The output of tin and arsenic also rose during the war years. It seemed as if good times had returned to the mining industry but there was a negative aspect to all this activity. Manpower was in short supply in all industries and mining was no exception. While women took over the work of men in all kinds of factories and manufacturing industries, this never applied to mining. They had never worked underground, but a few were still retained at surface work. As a result, every effort was concentrated on getting ore. Where in the past, a large proportion of the workforce would have been employed in creating shafts and levels to explore for new workable lodes, now everyone was involved purely in mining for ore. Lodes were being depleted with no new lodes being worked as replacements. This was to cause problems when the war ended.

In the immediate aftermath of the war, tin prices soared, reaching unprecedented levels of over £400 a ton. But by 1920, the price began to fall and what started as a slide soon became an avalanche. By 1922, the price had dropped to £140 a ton, while at the same time coal prices had more than trebled. Those mines that had not electrified were particularly hard hit, and some were forced to close; others carrying on pumping in the hope of better times ahead, others were simply abandoned. And it was not merely the lesser mines that suffered. Dolcoath, which had been one of the most productive of all mines and which, in 1920, was being worked down to a depth of 3,300 feet, closed down in 1921. A major factor in the failures was the way in which the mines had been worked to help the war effort; appeals were made to the government for help, but with no success. The coal mines were subsidised because coal supplies were essential for the economy. Tin prices were low and Cornwall now only supplied a small portion of the tin on sale worldwide.

This was a period of great distress in the mining districts. It was not just that miners were unemployed, but they no longer had the money to pay for the goods and services on offer by local tradesmen. The situation was so grim in Camborne that with no one having enough money to pay the rates, the street lighting had to be turned off. The means-tested dole was bitterly resented and inadequate and the families often had to rely on money being sent back to Cornwall by Cousin Jacks. Many simply left the area altogether to try and find work, either elsewhere in Britain, or in the minefields around the world. In 1900, there were more Cornish miners working abroad than there were in their native county.

The recession was short lived. In 1923, there was a rise in the price of tin and the government finally recognised that the metal mines needed some sort of support much as the coal mines did. Loans were offered with capital and interest guaranteed to the bigger companies under the Trade Facilities Acts of

1921 and 1922. The Acts were designed specifically to provide funds to relieve unemployment, provided the industries were considered basically viable. There was an injection of capital that was used to open up new ventures, while the mines that had continued pumping through the bad years were able to go straight back into production. Those that had flooded out faced a quite different future and many would remain closed

Attempts were made to form a cartel to keep the prices of tin high by buying in excess ore and stockpiling it until times improved. It was not an entirely successful venture and Cornwall was certainly not immune from the worldwide slump following the Wall Street crash of the 1930s. Once again, mines were closed and men found themselves out on the streets and it was probably little consolation to know that they shared their predicament with other workers in other industries. There was great publicity for the Jarrow marchers, who had lost their jobs when Palmers shipyard closed down. They walked to London to put the case for establishing new industries in the area and although they received great support from the general public, nothing significant ever materialised to help them. At least their case was widely publicised, but for the British population as whole, Cornwall was the county of sunshine, sea and sand, a holiday destination, not a centre of desperate hardship.

Unemployment ceased to be a problem with the outbreak of war in 1939, with young men being called up for service and with the demand for raw materials from local sources now crucial, the last four productive mines were kept busy producing essential metals. In the post-war period, new technologies made mining more efficient. In the past, the only way to discover new lodes had been via cross cuts and shafts, but the introduction of diamond drills made that unnecessary to some extent. The drill is a tube with diamonds set round the rim. The head is rotated, cutting out a core that eventually fills the pipe. This can then be removed and studied. As the drill goes deeper into the rock, so a collection of cores is made that provides a cross section of the ground. Greater efficiency was one factor in favour of the new mines, but they were faced by increasing competition, especially from the Far East, where costs per ton labour were far lower, due in part to lower wages. In 1951, the First International Tin Agreement was signed, designed to regulate the prices and keep them high. It worked as the old cartel system had, with tin being stockpiled. The International Tin Council was set up and was soon buying surplus stock and arranging loans with banks around the world. The cartel included both producers and consumers of tin, each country's vote being weighted by the size of its exports and imports.

It seemed that the Cornish industry had an assured future, and mines were investing in modernisation on a large scale. At Geevor mine, the Victory shaft had been sunk in 1919 and in 1954, the old wooden headgear was replaced by the steel headgear that can still be seen today, with electric power replacing steam for winding. By 1975, it had reached a depth of 1,575 feet and it was decided to extend it by means of an incline that would connect with the old Levant workings under the sea. The incline reached a depth of 2,130 feet and electric power was supplied underground to work a conveyor belt that carried material from the working face to the foot of the Victory shaft and to operate a man train for the miners. It was considered such a major development that the Queen came to Cornwall in 1980 for the official opening of the incline. At Crofty, the men used to ride on the belt on the incliner below the 380 fathom level – a mode of transport I've used in coal mines, which makes up for in speed what it lacks in comfort.

In the early 1980s, I was working on a series of programmes for BBC Radio 4, with an accompanying book, called *Britain Revisited* in which I followed in the footsteps of earlier writers. One of these was Wilkie Collins and an essential element in that story was his mine visit, so I was able to arrange a trip down Geevor. Collins had been kitted out in miners' clothes that were far too big for him, and my own clothing was scarcely less incongruous. I was given a set of dungarees that only fastened down one side and flapped open on the other, a dust coat that might have been effective had it had any means of fastening and a pair of rubber boots. Unlike Collins, I did not have to rely on a candle stuck in a lump of clay to light my passage, but was equipped with the familiar hard hat and battery lamp. The descent in the cage was rapid but familiar from many previous visits to collieries, but the incline was certainly new and the first impression was of an impressively modern and prosperous operation.

Having reached the lower levels, it was a surprise to discover passageways lit by strip lighting, wide and tall enough that one could walk comfortably, unlike most collieries I had visited where, being inexperienced, my hard hat would beat a regular tattoo on the roof. That wide, well-lit section eventually ended and the going got a little less comfortable. Slopping through ankle-deep water. It was a sobering thought that to keep things this dry, two underground pumps were needed; one pumping out 160,000 gallons of sea water every day, the other removing 360,000 gallons of fresh water. There were four classes of workers underground: trammers; diggers; timber men; and day pay men. The trammers were the first we met, emptying ore over a grizzly. On we splashed through orange-tinted, greasy water until there was a distant rumble that gradually got louder until we turned a corner and there was the rock wall and two spectral figures in a halo of light, drilling holes ready for blasting. They were advancing at the impressive rate of 8 or 9 feet a day and blasting out some 25 tons of granite. They were the day pay men. I had half expected to hear as Collins had the rumble of waves overhead, but we were far too deep for any such noise. The return to the surface was via an older shaft, full of slight bends – the men called it 'going up the banana' and it was only when I had to make the long walk back to the changing rooms that I was aware just how far I had travelled underground. It had all been most impressive, with no hint that it would all too soon come to an end.

The Tin Agreement only lasted on and off for about thirty years. Without it, Geevor and Crofty could never have survived. In 1985, it was revealed that the Council had debts of £900 million and no means of repaying them. The result was a drastic collapse in tin prices and the effects reverberated through the mining communities everywhere. Cornwall's industry was devastated. Geevor closed down in 1991; the pumps were shut down and the works were flooded. It was not the end of the story. Shortly after closure, the site was developed as an important heritage centre and opened to the general public. Wheal Jane, Cornwall's most modern mine, was opened in the 1970s and closed in 1991. The other important, modernised mine was South Crofty and that remained open right up to 1998 when it too succumbed to financial pressures and all work came to an end. In the twenty-first century, tin production has mainly been concentrated in the Far East. Today, China is the major producer with an output of 125,000 tonnes a year, closely followed by Indonesia with 84,000 tonnes. The price of tin has soared again, and whenever that happens, there is always speculation that Cornish mining might restart. Perhaps there is still another chapter to write in the long history of metal mining in the South West of England

Chapter 17

PLACES TO VISIT

Although commercial mining has ceased, even if only temporarily, there are still a number of sites where one can see the processes, buildings and machines of this once great industry. It is one thing, for example, to read about Cornish beam engines, but nothing compares with the actual sight. Even film of a working engine cannot capture the whole experience and certainly can't reproduce the unique smell of a working engine. And an underground visit gives at least some idea of the working conditions – even if nowadays no one will expect you to climb up and down vertiginous ladders. The following descriptions will, I hope, tempt readers to explore these sites for themselves.

There are, of course, many more sites that can be explored, especially the many engine houses that have been abandoned and simply left to gently decay. The sites listed here are all ones that have been specially preserved for the general public.

BLUE HILLS TIN STREAMS

Trevellas Coombe, St. Agnes TR5 0YW

This is a most romantic site in a beautiful, remote deep valley. The ore that is worked here is collected from the shore beneath the cliffs, where it has been washed out by the waves. It is taken back to the site, where it is crushed by the stamps, powered by the water wheel. It is then smelted in a small furnace, and worked into jewellery by craftsmen on site.

BOTALLACK ENGINE HOUSES

Botallack, St. Just TR19 7QQ

The site is approached on a track from Botallack village to the coast. The Crown engine houses are the most photographed in Cornwall and rightly so. The lower Crown engine house perched on its rocky ledge with a background of breaking waves makes for an imposing sight. But there is more to the site than this. Before reaching the cliff top, the track passes the former count house and the extensive remains of the arsenic treatment plant, dominated by the tall stack that took the fumes away from the calcining furnace. At ground level, there is an elaborate system of flues and chambers, where the arsenic was collected. There are also remains of the tin dressing system and, from a later age, the ruins of the building that once held the compressors for the rock drills.

DANESCOMBE MINE

Calstock, Cornwall PL18 9SG

This is unlike the other sites listed in that, strictly speaking, it is not visitable in the ordinary way. But the engine house has been converted into a holiday home by the Landmark Trust and provides a rare chance to stay in an engine house, rather than simply look at it. It originally housed a 40-inch rotary engine and was used for pumping and working dressing machinery. It was operated for the Cotehele Consols copper and arsenic mine from 1822 to 1900. Information on lettings from www.landmarktrust.co.uk

EAST POOL ENGINE HOUSES

Pool Nr Redruth, Cornwall TR15 3NP

The whim engine house is a prominent feature right next to the main road through Pool. Inside is a beam engine, built by Holman's in 1887. It is typical of many whim engines that were once in use throughout the region. The main engine is inside the building, and the beam has its fulcrum on the bob wall. Outside, the beam is connected to a sweep arm and crank to turn the winding drum. The engine was originally worked by steam from a Cornish boiler, and although that is still in situ, alas it is no longer used. On demonstration days, the engine is turned over using an electric motor, but at the time of writing the whole engine is awaiting repairs, though it is hoped to have it moving again in the future.

On the far side of the main road is the magnificent pumping engine that stood over the Taylor shaft of East Pool and Agar Mine. This is a massive machine with a 90 inch diameter cylinder made by the Harvey Foundry in 1892, which remained working right up to 1955. The pumps brought up water from a depth of 290 fathoms below adit. Sadly, the engine is static, but it is hugely impressive and worth seeing just to understand the scale of operations. Fittingly, the engine room is graced by a bust of Richard Trevithick.

GEEVOR TIN MINE

Pendeen, Cornwall TR19 7EW

Geevor was one of the last mines to be opened in Cornwall, started in 1911, and one of the last to close in 1990. As a result it provides a unique opportunity to see state of the art mining technology of the twentieth century. The main shaft, the Victory, was originally worked by a steam winder and had a wooden headgear; today the headgear is steel and the winder electric. There is a chance to go underground, but this is in the earlier Wheal Mexico mine that was incorporated into Geevor, but was originally worked in the eighteenth century.

The surface features include the dry where the miners changed which has been left exactly as they were when the mine closed. The dressing mill has also been preserved. The museum area includes part of the Holman collection, originally housed in its own museum in Camborne. It contains examples of the many different kinds of machinery produced in this famous works over the years.

KING EDWARD MINE MUSEUM

Troon, Camborne, Cornwall TR14 9HW

The setting up of the mine as part of the Camborne School of Mines was described in Chapter 15. When operations were moved away from Camborne, the machinery at the mine was mainly scrapped. However, volunteers were able to salvage appropriate fixtures from other mine closures and recreate the old mill building. So visitors today can see such imposing machines as the California stamps, the round frame and rag frame. When the volunteers discovered the parts of the round frame for reassembly they had all been numbered – but there was no key to what those numbers represented. Nevertheless, the whole assembly was completed and put back into working order. Rag frames were first introduced to Cornwall in the 1860s to recover the last traces of tin from the mill. They consisted of sloping wooden benches down which the material was pushed by regular cascades of water that carried the heavy tin down the slope to be separated out. No rag frames had survived, but Willie Uren, a retired mill foreman, built one from memory. All the machinery is demonstrated.

In 1908, the School purchased a new Holman steam winder that was geared to work two drums. This is now worked using compressed air and provides an excellent simulation of how it would have operated under steam pressure. The former boiler house now houses an exhibition, with material from Holman's, including their famous rock drills.

LEVANT MINE

Trewellard, Pendeen, Nr. St. Just, Cornwall TR19 7SX

Like nearby Botallack, Levant mine occupies a dramatic clifftop site, and it is still possible to peer down the deep cleft on the rocks to see the steps down which the miners made their way to enter the adit. Unlike the Botallack engine houses, Levant still has its steam engine – the last in the region that is still worked by steam. The engine is a whim engine and comparatively small with a 24 inch cylinder, built by Harveys in 1840 and rebuilt in 1862 following an accident. Unusually, the whole engine is inside the house. The drive is taken to the drum outside the engine house that would originally have been connected to the headgear above the shaft. The headgear remains and visitors can make themselves giddy by staring down into the shaft. Outside the buildings are the remains of the man engines, left there following the tragedy of 1919. The whole site is now in the care of the National Trust.

MORWELLHAM QUAY

Tavistock, Devon PL19 8JL

This was once a major port for the shipping of copper ore, connected to the mining area near Tavistock by the Tavistock Canal. This can be seen emerging from a tunnel on the hillside above the quay. Wagons would then have been lowered down the incline and carried on the raised tramway to the waiting boats. One of these vessels, the ketch *Garlandstone*, can be seen moored at the docks. Built in 1909 at James Goss's yard at Calstock, it would have carried ore from the Tamar round the coast to South Wales for smelting. A similar ketch *Shamrock* built at Plymouth in 1899 can be seen at Cotehele Quay and has been fully restored to sailing condition. She had a rather less glamorous cargo to carry in her working days – she was a manure barge. The other outstanding feature on the quay is the water wheel that once powered the ore crusher. Many of the buildings have been restored for use as craft centres and the inn and brewery provide a welcome for visitors.

A tramway from the quay leads to the copper and arsenic mine, George and Charlotte, which was worked from the early 1700s until the middle of the nineteenth century. Access to the mine was always by adit rather than shaft, and today visitors are taken into the workings on a small train. One unique survivor is the underground water wheel.

POLDARK MINE

Trenear, Helston, Cornwall TR13 0ES

The name has obviously been chosen to attract visitors who enjoy the Poldark books and TV series, but there is nothing fictional about the site. This was originally the Wheal Roots and was worked for tin enjoying great prosperity in the eighteenth century. Visitors have a choice of mine tours, which take them down to either three or four levels. One fascinating feature of the underground tour is the shaft with the infamous ladder that had to be climbed by the working miners – the visitors are not surprisingly asked to use it.

Above ground, there is a fascinating collection of machinery, not all directly connected with the mining industry, but all of interest. A 30 inch beam engine by Harvey's of Hayle, built in 1846 and originally used at the Bunny tin mine, is a prominent attraction. There are two air compressors, one originally powered by steam, the other by electricity. An unusual exhibit is the ⅙ model of a traversing winding engine, exhibited at the Paris World Trade Fair of 1900 by Holman. The alternative to either steam or electric power is represented by two Tangye pumps, worked by an oil engine. In the days before the establishment of the national grid, a company wanting electricity had to build its own generating plant. The power house at Newton Abbot was home to a magnificent triple expansion engine that drove a generator producing 400 KW DC. An early Holman rock drill of 1882 is on display. There are other exhibits of all kinds, including locomotives.

TOLGUS TIN

New Portreath Road, Nr. Redruth, Cornwall TR16 4HN

This is a tin streaming works, where the ore is extracted from the stream running through the site. The mill was run by the Stewart family from the middle of the nineteenth century up to 1968. The mill contains nineteenth century stamps and separating machinery. The ore is smelted on site and used to make jewellery.

WHEAL BETSY

Off the A386 north of Mary Tavy, Devon PL19 9PB

The silver and lead mine was reopened in 1806 and was water-powered until 1868, when steam was introduced – but the mine closed just eleven years later. The engine house still stands in a little valley and is now in the care of the National Trust. It is a reminder of what was once an important Devon industry.

THOMAS NEWCOMEN ENGINE

The Engine House, Mayor's Avenue, Dartmouth, Devon TQ6 9YY

The Newcomen pumping engine transformed mining in the early eighteenth century and this is a rare survivor. It was originally built c.1725 and served two collieries before being moved to Hawkesbury Junction in 1821 to supply water to the Coventry Canal. It remained in use right up to 1913 and the original engine house can still be seen beside the canal. In 1963, it was moved to Newcomen's home town and housed in a specially built Engine House. With its wooden beam, the chain connecting the beam to the piston rod in the open-topped cylinder, it looks very different from the familiar beam engine of the Watt era.

SELECT BIBLIOGRAPHY

Agricola, *De Re Metallica,* 1556

Barton, D.B., *The Cornish Beam Engine,* Wheaton, 1967

Barton, D.B., *A History of Tin Mining and Smelting in Cornwall,* 1989

Booker, Frank, *The Industrial Archaeology of the Tamar Valley,* David & Charles, 1967

Brooks, Tony and Watton, John, *King Edward Mine: An Illustrated Account of Underground and Surface Operations 1897 - 2001,* Cornish Hillside Publications, 2002

Burt, Roger, *Cornish Mining,* 1969

Burton, Anthony, *Richard Trevithick,* Aurum, 2002

Carew, Richard, *Survey of Cornwall,* 1602

Earl, Bryan, *Cornish Mining: The Techniques of Metal Mining in the West of England, Past and Present,* Cornish Hillside Publications, 1968

Hamilton Jenkins, A.K., *The Cornish Miner,* 1927

Hocking, Anthony, *South African Mining,* 1975

Lovell, Langford, *West Barbary,* 1891

Payton, Philip J. *The Cornish Miner in Australia: Cousin Jack Down Under,* Truran, 1984

Pryce, William, *Mineralogia Cornubiensis,* 1778

Randall Lewis, G., *The Stannaries,* 1908

Rowse, A.L., *The Cornish in America,* Macmillan, 1959

Todd, A.C., *The Cornish Miner in America,* 1967

A.C. Todd and Peter Laws, *The Industrial Archaeology of Cornwall,* David & Charles, 1972

Edmund Vale, *The Harveys of Hayle,* 1966

INDEX